T0324941

Heavy Equipment Operation and Maintenance Manual

Starting from the purchase of heavy equipment and following through to the end of its economic life, this manual explains how to efficiently maintain and operate different types of heavy equipment. Assigning heavy equipment to different projects and utilizing them in varied systems is a large part of construction operation; ensuring everything is monitored consistently and maintained accordingly is essential. This book aids engineers in facilitating straightforward, consistent reporting systems and cost-efficient equipment use. This is particularly of note to the construction industry.

Features:

- Enables engineers to save time and money on maintenance costs and maximize the availability of the heavy equipment
- Provides comprehensive coverage of methods and procedures for heavy equipment management
- Provides charts for practical use by engineers in the field, e.g., mapping out a maintenance schedule
- Includes chapters on maintenance and field operations organization, including safety and security organization

This book will be of interest to construction engineers, plant engineers, mechanical engineers, maintenance plant and field engineers.

Heavy Equipment Operation and Maintenance Manual

Ernesto A. Guzman

CRC Press
Taylor & Francis Group
Boca Raton London New York

CRC Press is an imprint of the
Taylor & Francis Group, an **informa** business

Designed cover image: Ernesto A. Guzman

First edition published 2024
by CRC Press
2385 Executive Center Drive, Suite 320, Boca Raton, FL 33431

and by CRC Press
4 Park Square, Milton Park, Abingdon, Oxon, OX14 4RN

CRC Press is an imprint of Taylor & Francis Group, LLC

© 2024 Ernesto A. Guzman

Library of Congress Cataloging-in-Publication Data

Names: Guzmán, Ernesto, 1971- author.
Title: Heavy equipment operation and maintenance manual / Ernesto A. Guzman.
Description: First edition. I Boca Raton, FL : CRC Press, [2024] I Includes
bibliographical references and index. I Identifiers: LCCN 2023011560 (print) I LCCN
2023011561 (ebook) I ISBN 9781032419800 (hbk) I ISBN 9781032419817 (pbk) I
ISBN 9781003360667 (ebk)
Subjects: LCSH: Construction equipment--Handbooks, manuals, etc.
Classification: LCC TH900 .G885 2024 (print) I LCC TH900 (ebook) I DDC
624.028/4--dc23/eng/20230427
LC record available at https://lccn.loc.gov/2023011560
LC ebook record available at https://lccn.loc.gov/2023011561

ISBN: 9781032419800 (hbk)
ISBN: 9781032419817 (pbk)
ISBN: 9781003360667 (ebk)

DOI: 10.1201/9781003360667

Typeset in Times
by KnowledgeWorks Global Ltd.

Support material available at https://resourcecentre.routledge.com/books/9781032419800

Contents

PART IV *Project Operations*

PART V *Process Plants*

PART VI *Equipment Safety and Security*

PART VII *Equipment Rental Rate*

About the Author

 From 1984 to 2013, **Ernesto A. Guzman** commuted to Jakarta, Indonesia, and Hail and Jeddah, Saudi Arabia, from Manila, Philippines, for much of his professional and academic concerns. Being a mechanical engineer, he worked as a heavy equipment expert for twelve years in the Philippines and then continued his path as an expat in Indonesia and Saudi Arabia for another twelve years on. He has been involved with construction projects at Philippine National Construction Corporation and general manager of Tierra Factor Corporation, both in the Philippines; Assistant Director for PT Bukaka Teknik Utama (Indonesia) for the Java steel tower transmission lines project; and general manager for PT Surya Mahkota Timber (Indonesia) for furniture manufacturing and project manager of Pt Eurasia Wood Industries (Indonesia) MDF-wood furniture manufacturing operations in Jambi, Sumatra, and Jakarta, Indonesia, among others.

In addition, he was also consultant to PT Sulmil Wood Industry (Sulawesi, Indonesia), Australia Consolidated Industries Insulation (Phil). And at Al Khodari and Sons, North Rail Project at Al Jouf, Saudi Arabia. He also taught at the University of Santo Tomas Graduate School (Philippines) lecturing on Production Management and Supply Chain Management. Then, he proceeded to lecture at the Royal Melbourne Institute of Technology (RMIT) International University in Ho Chi Ming City, Vietnam, also handling Supply Chain Management and Business Statistics.

He has written many papers, articles and manuals in various subjects from operations and maintenance of heavy construction equipment, quality and test procedure to methods of construction and quality assurance, material inspection and testing, including the article Industrialization for Development about the industrialization path for economic development of a nation, and on the light side, he has written the Ying and Yang in Contemporary Times.

He earned his Bachelor of Science in Mechanical Engineering at the University of Santo Tomas (UST) and his Master in Management degree at the Asian Institute of Management (AIM), both in the Philippines. His thesis was Corporate Strategic for Philippine National Construction Corporation (PNCC) Tollway Division. A study for the expansion of two main expressways, the North Luzon Expressway (NLEX), 83 km length, and the South Luzon Expressway (SLEX), 43 km length. He completed his academics for his PhD in Technology Management at the Technology University of the Philippine in Manila.

Introduction

Heavy equipment management is a major component of the construction industry. It can account alone for 57 percent of the total assent of a company. An equipment-proficient management system is paramount in order to be effective and profit-driven as a principal item of the corporation in the construction industry. Construction is a prime mover industry. It has been known to help the economy of a country sustain its economic growth during the period project work is in progress and after the project has been completed. In the construction business, the best equipment-managed company can be assured to have the most out of their heavy equipment.

An equipment management manual is one of the most tried and tested reference to attain this. Cost in equipment operations are most likely to end up with hidden and arbitrary cost left unidentified due to lack of structure and procedures in its operations. This can devour profit without the company board knowing what has happened. The best way is to have standard equipment procedures to minimize and eliminate wasteful operation practices.

The manual is a collection of heavy equipment working procedures that cover the equipment department organization, administration, maintenance and field operations of the equipment. From the period the equipment was purchased, assigned it first field project work and moved project to project up to the time the equipment shall have reached its economic life. The manual was prepared for ownership of at least 10–100 or more equipment in the fleet of a construction company.

It has been written to ensure the equipment technical and field personnel will clearly understand key procedures, information and technical concepts from the base level so as to apply and execute the procedures with ease and accuracy, leaving no room for error and misunderstanding, especially in the construction operations field where no indecision is tolerated.

Every chapter has a corresponding chart or form to accomplish the proper procedure outlined in the chapter. Each chart gives direct, factual and concise data to ensure the procedure is executed with accuracy. Each form will contain the exact data for each specific operation so as to provide the exact information for the engineer and manager on hand as well as for the equipment management officials in their command offices.

The number of these vital documents for equipment operation and maintenance (Figure 0.1):

- Fifteen (15) Charts and Tables.
- Twenty-Seven (27) Forms from equipment coding, manpower classification, branding, inspection, operations, PM schedules, time and cost, rental rate, safety and security coverage.

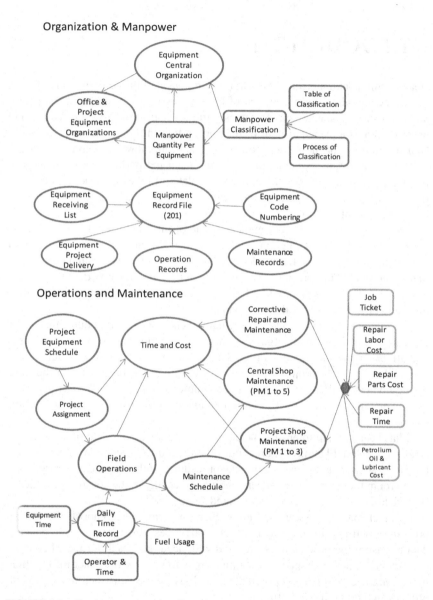

FIGURE 0.1   Equipment operation flow chart.

This book is the product of 20 years of work experience in the construction field as equipment manager to senior manager of construction companies that have completed projects such as hundreds of kilometers of expressways and roads, bridges, sugar mills, fertilizer manufacturing plants among other construction projects in the Philippines, Saudi Arabia, Indonesia and Malaysia.

Part I

Equipment Organization

1 Equipment Management Organization and Administration

1.1 GENERAL ORGANIZATION AND GENERAL FLOW CHART

One of the first important stages in equipment management is to have a well-organized organizational structure. There are two main structures, the centralized general organization where the central shop and yard and main equipment office are located and the satellite equipment project organization.

These tables of organization focus on policies and systems instead of people. The personnel assigned for each position have a distinctive function to avert any duplication. There is also a strong command responsibility of each from the manager to superintendents, to supervisor to engineers and down to the workforce of operators and maintenance crew.

Teamwork, collaboration and communication make the organization active and busy, when the projects are in operation. Equipment team members can rely on each other for guidance and support. This enables them to focus on the equipment operations' goals and carry out their duties and responsibilities effectively.

With a well-organized table of organization, the company can have as many as 10–15 different projects operating at the same time and located in different locations. The well-defined functions and positions will organize the equipment project operation efficiently and effectively.

Equipment Management Department shall be the sole responsible for all the company equipment, and this responsibility shall include administration, operations and maintenance functions of the equipment.

More specifically, these are the management and technical functions for purchasing, shipping, utilization, maintenance, repair, security, safety and disposal of equipment. Also included are the spare and repair parts sourcing, inventory and warehousing, central yard maintenance, central shop facilities and technical services.

To facilitate these functions, there are eight basic sections operating under the department. These are the central shop and yard, the technical services, field operations, personnel and administration, finance, accounting, purchasing and warehouse and inventory control. If it so warrant, another section can be separated, the transport and shipping section.

For some operations, a process plant can also be attached to the department. A department can be centralized or each project equipment operation can be created, according to the convenience of the projects locations an duration.

DOI: 10.1201/9781003360667-2

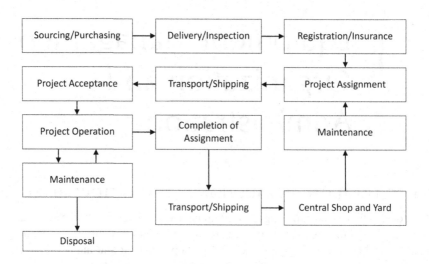

FIGURE 1.1   Equipment ownership flow chart.

A project can also be a centralized and a standalone if the project is large enough and the construction period long enough where equipment will be fully depreciated at the end of the project. Whichever way, a well-organized set is mandatory.

The flow chart of a simplified equipment ownership and operation of a construction company is shown in Figure 1.1.

2 Equipment Tables of Organization

2.1 CENTRALIZED ORGANIZATION

A company with a centralized equipment operation shall have all the functions of an independent organization. This shall act as the head office of equipment operation. It shall be composed of two major functions, the main functions which are the central shop and yard, the technical services and the field operations and the support functions which are the purchasing, warehouse, inventory control, personnel and transport.

If the equipment operation is a profit center, the finance with the cashier and the accounting shall be part of the equipment division. If the equipment operation is a cost center, these two functions will be located at the construction central office, but bookkeeping and a cashier will be part of the equipment central office.

Another option for a cost center operation is to locate the purchasing function with the central or head office of the company. But the warehouse and inventory control will be located with the central shop and yard location.

Whether it is a cost center or profit center, when a fleet equipment is deployed in a far out and separate project location, a project equipment organization shall be set up. This shall be called the equipment project section. A small set of technical people shall compose this project equipment field office. They shall be reporting both administratively to the equipment head office and operation wise to the project office they belong.

This is the reason documentation and very comprehensible organization functions be set up. While it may look simple, during actual project operation period, confusion can result if an unsystematic organization is set up adding to high hidden cost to the company (Figure 2.1).

2.2 PROJECT EQUIPMENT ORGANIZATION

The project equipment organization is a satellite unit of the equipment department located at the job site. It is designed to do field equipment operations and maintenance. It shall have a small warehouse and petroleum, oil, and lubricants (POL) distributor depot. A time and cost unit will carry out the equipment monitoring, and it shall submit all equipment time and cost reports to the technical services at the central office every month.

DOI: 10.1201/9781003360667-3

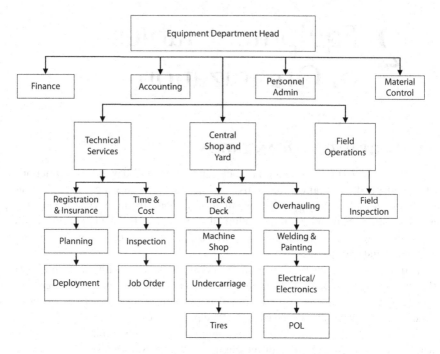

FIGURE 2.1   Centralized general table of organization.

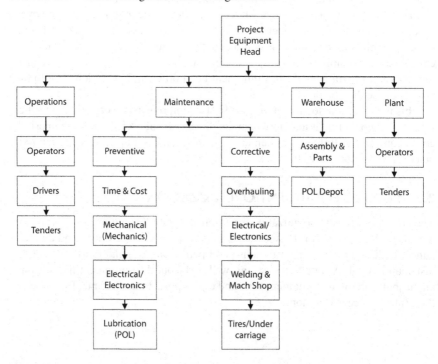

FIGURE 2.2   Typical project equipment table of organization.

This in turn shall be used to consolidate all the equipment utilization and availability reports for the management. The number of personnel in each set up shall depend on the number of equipment deployed in the project. This is shown in Section 3.4 in Chapter 3. The project equipment office is headed by a representative of the equipment head office. He is usually an Equipment Superintendent (Figure 2.2).

3 Equipment Management Personnel and Manpower

3.1 MANPOWER SKILLS AND CLASSIFICATION

Personnel and manpower classification for equipment spells the difference in the management of equipment, especially if projects are situated at different and remote locations. People who will execute general plans, movement and maintenance of the equipment fleet are the conduits of management to an effective and efficient fixed assets administration. The profit and return on investment shall be dictated by the work harmony of the central equipment organization and the different units established by management.

A construction company established to do more than one project will benefit from a good personnel and manpower contingent even though equipment is transferred and utilized from one project to another. Presently, personnel and manpower is designed not any more as human resources but as human capital. This is to give emphasis on how personnel and manpower are indeed part of the assets of a company, especially after having them learned and be skillful on their respective jobs.

They become contributor to cost savings for the company. Cost is minimized and equipment is well maintained when the personnel and manpower do their share of efficient and effective operation. The secret lies in good system of personnel and manpower quality and quantity selection for the company to make them profit-contributing factor in the company.

It is mandate of the company to have good skill qualification details and testing procedures. Having these, the company should know the minimum and effective number of skills it needs for its operation and maintenance. Here are the classifications and the ultimate quantity for a given number of equipment of a company.

3.1.1 STAFF AND MANPOWER (PARTIAL LIST) CLASSIFICATION

Complete list of personnel and manpower is based on the table of organization (Figures 3.1 and 3.2).

DOI: 10.1201/9781003360667-4

FIGURE 3.1   Wheel loader.

FIGURE 3.2   Dump truck and wheel loader.

Staff and Manpower List

Staff	Officers
1. Equipment Head	VP/AVP/Senior Manager
2. Technical Services/Central Shop & Yard Head	Senior Manager/Manager
3. Shop Head/Transport Head	Superintendent
4. Section Head	Supervisor/Senior Engineer
5. Engineer Staff	Engineer I, II, III
6. Property Staff	Property Custodian I, II, III
7. Clerical Staff	Clerk I, II, III
8. Equipment Inspector	Inspector I, II, III

Field Operation Ranking	Maintenance Ranking
1. Dozer Operator I, II, III	1. Lead Man
2. Wheel Loader Operator I, II, III	2. Senior Mechanic/Electrician/etc.
3. Grader Operator I, II, III	3. Mechanic I, II, III
4. Crane Operator I,II,III	4. Electrician/Electronics Tech I,II, III
5. Driver I, II, III	5. Machinist I, II, III
6. Tender I, II, III (Generator, Compressor)	6. Welder I, II, III
7. Light Operator I, II, III	7. Painter I, II, III
8. Trailer Tractor Driver I,II, III	8. Tire Man I, II, III
9. Roller Operator I, II, III	9. Lube/Oiler Man I, II, III
10. Plant Operator I, II, III	10. Handy Man

3.2 MANPOWER SKILL CLASSIFICATION AND UPGRADE

This classification will not only tell apart different skills but also separate those with basic skill from those with long years of experience and good performance. The benefit therefore can be deemed from the following purposes:

1. To unify system of employment classification for skilled and semi-skilled personnel.
2. To upgrade classification system to allow longtime personnel to be re-evaluated and be updated into the new system and in par with new hires.
3. To be used for classification of hence on new personnel.
4. To be used annually to evaluate personnel according to his yearly performance of his assignment in the company.
5. To establish the record of performance of each personnel including each accomplishment and commendable work contributions as well as unhelpful and unsafe performances.
6. To be ready for International Organization for Standardization (ISO) classification for the prospective development of the company in the future.

3.2.1 PERSONNEL CLASSIFICATION

Each personnel has to be classified according to the criterion enumerated below and/or as the equipment administration shall so approve.

TABLE 3.1
Table of Classifications

Classification	Definition
One (I)	Entry classification or according to criterion set in Table 3.1 or 3.2. This rank is also for those with basic skills proficiency.
Two (II)	With years of experience per Tables 3.1 and 3.2. Skills shall be classified as good and can be trusted in his work assignment.
Three (III)	With years of experience per Tables 3.1 and 3.2. Skills shall be classified as very good and can work with minimum of supervision.

Examples:

Mechanic 1	- Entry level or per classification, see the tables below.
Mechanic II	- With some years of experience with other company and has good knowledge of work assignment.
Mechanic III	- With many years of experience and have been with company for long and knows the job excellently.
Electrician I	- Entry level or per classification above.
Service Driver I	- Entry level or per classification above.
Trailer Driver III	- Long time with company and is expert in his work assignment.
Welder II	- (See above)
Dozer Operator II	- (See above)
Crane Operator III	- (See above)

Note: The same goes on for all other positions starting from I to III (refer to the Table of Organization).

3.2.2 APPLICATION

To apply and implement this manpower classification scheme when the company manpower is yet unclassified, enumerated below are two steps on how to adapt the scheme. First is for the current employees of the equipment department, and the other is for those to be hired by the company. For implementations, there are two suggested options below.

3.2.2.1 For Current Employees

Create a performance sheet for every manpower skill for each current employee to evaluate performance and tenure.

Test and evaluate the performance and skill level of the personnel and enter to their 201 file or database as one component of their evaluation.

Evaluate their year of service in the company and enter in their performance sheet.

3.2.2.2 For New Hires

Classify the new hires immediately and as per policy, according to the criteria they shall qualify, use performance sheet given in item A above.

After their classification, they shall be on line to the policy annual update for their classification and job renewal in the company throughout their tenure.

3.2.3 PROCEDURES FOR IMPLEMENTATION

3.2.3.1 For Current Employees

- **Option 1**
  - Establish the date of hire of each employee. This shall be used for each, his re-classification if needed and hence his evaluation of performance each year to renew his contract or be promoted to the next level whenever appropriate.
  - Date of evaluation shall be prepared one (1) month before the date of hire or anniversary date of the employee.
  - Each project shall, through the Project Manager and Equipment Manager, evaluate the work performance of each employee on his date of hire or anniversary.
  - A criterion for this evaluation shall be established; see sample Table 3.1.
  - Evaluations shall be on the job during the employee's working time at the field or shop or place of work, as the case maybe and with his equipment, machine, tools or labor.

TABLE 3.2
Evaluation Criterion for Current Employees

Skills Evaluation

Classification	Years of Experience with Company (A)	Points	Equipment Knowhow Evaluation (B)	Points	Accident Record (C)
One (1)	3–5 years	1	Basic	1	More than one accident
Two (2)	6–10 years	2	Good	2	One accident
Three (3)	11 years on	3	Very good	3	None

Personnel Evaluation

Classification	Attitude (D)	Points	Cooperation (E)	Points	Efficiency (F)	Points
One (I)	Basic	1	Basic	1	Basic	1
Two (II)	Good	2	Good	2	Good	2
Three (III)	Very good	3	Very good	3	Very good	3

The result shall then be submitted to the Central Office Administration Division for audit, accountability, verification and final approval.

- **Option 2**
  - Prepare a team of evaluators with one Equipment Supervisor and Administration Officer each from the central office to join and work with the project Equipment and Fleet Manager/s and Maintenance Manager with their assigned supervisors to evaluate their project-assigned personnel with a prescribed period set by Administration Division.
  - Evaluation shall be on the job as the employee performs his daily work assignment or when needed a special task to determine his skills.
  - The result shall be compiled, summarized and approved by the Project Manager and Equipment Manager.
  - This shall then be submitted to the Central Office Administration Division for audit, accountability, verification and final approval.

3.2.4 FINAL EVALUATION

- Add $((A+B+C+D+E+F)/6)$ and this is the average point of evaluation.
- Employee based on criterion above shall determine his rank.
- Establish classification and record for each employee according to his points garnered in the evaluation.

3.2.4.1 For New Employees

Establish the date of hire of each employee. This shall be used as his anniversary date or date of annual evaluation of his work performance.

The employee shall be tested and classified at the central yard prior to his assignment to the project site.

The Equipment Central Yard shall submit the classification to the Administration Division for approval prior to his assignment to the any company project site or office.

The new employee shall have their classification, salary level and employment status. This shall be evaluated on their next anniversary date wherever they may be assigned (Table 3.3).

TABLE 3.3

Evaluation Criterion for New Employees

			Skills Evaluations			
Classification	**Years of Experience (A)**	**Points**	**Equipment Knowhow Evaluation (B)**	**Points**	**Field Work Evaluation (C)**	
One (I)	1–7 years	1	Basic	1	More than 1 accident	
Two (II)	7–12 years	2	Good	2	One accident	
Three (III)	11 years on	3	Very good	3	None	

3.2.5 FINAL EVALUATION

Add and take the average point of each employee based on criterion above.

Classify each employee according to his points garnered in his evaluation (same as Table 3.1).

Other evaluation criterion for the operator/driver written exams maybe included.

3.3 MANPOWER REPORT

3.3.1 MONTHLY EQUIPMENT MANPOWER ROSTER

A monthly equipment manpower roster shall be prepared by all section and units in the table of organization, both in the central office and the project office. This is to account all personnel and manpower in each project of the company. This report shall be submitted to the Technical Services department who in turn shall consolidate and summarize the reports. The equipment central office report shall be first a detailed of names and classification or position, and then a summary of quantity of each classification or position.

For the equipment project summary report, first is the project equipment manpower report (see sample below), then the consolidated and quantity report of each project (see below). The central office may further consolidate the report into one equipment month report. However, the central office and the project reports can be analyzed individually according to their operations.

First Report: The Equipment Manpower of XXXX Project:

(Sample Manpower Report)

Form No._____

Company Name & Address

First Report: Typical Equipment, Manpower XXXX
Project Report
Project: XXX Railway Project
Location: Riyadh, Saudi Arabia

Shift: Morning Shift

Month: Xx January 2020

No.	ID Code No.	Name	Position	Nationality	Joining Date	Equipment No.
1	501	IHAB ABDU	SUPERINTENDENT	JORDAN	1/11/2005	Office
2	7923	SWRAPOL SAWNPEE	SUPERVISOR	NEPAL	28/3/06	Office
3	12040	HABIBU RAMAN	ENGINEER III	PAKISTAN	12/3/2008	Office
4	12030	AURAZEB KHAN	SENIOR ENGINEER	PAKISTAN	ON VAC.	Office
175	15778	RAJEEVA ABDUL	MECHANIC III	INDIAN	21/1/2010	Shop
176	16189	ISLAMKAJI DOM	MECHANIC III	INDIAN	VACATION	Shop
177	17656	CHANDRA YUNG	SVT DRIVER III	INDIAN	25/12/2007	SVT 66
178	18390	TAWAREE AHMED	WL OPERATOR II	INDIAN		WL 50
179	2011	REAAN FEFTER	GRADER OPER.	EGYPT	ON VAC.	MG 41
200	32535	ANTHONY LAJORE	TrT Driver I	PHILIPPINE	18/12/2005	TrT 82
201	32562	NISAN SULAIMA	SV DRIVER II	PHILIPPINE	2/12/2004	SV 63
202	32548	RUEL REYES	CT OPERATOR I	PHILIPPINE	22/12/2005	CT 59
604	45243	MOTILAL DEEN	CT OPERATOR III	INDIAN	9/4/2006	CT 71
605	45278	RAKEESF KUMAR	CT OPERATOR I	INDIAN	9/4/2005	CT 106
606	45254	MOHD ITERN	TIRE MAN	INDIAN	9/4/2004	Shop
607	45723	ALAUDDIN POSH	TT DRIVER II	INDIAN	25/12/2007	TT 74
608	52878	ABDUL HANNA	COOK	BANGLADESH	1/12/2004	Mess Hall
609	53033	KADIR MASHEEN	TENDER	BANGLADESH	12/12/2008	Shop

Note:
- Separated by nationality, this can be re-sort per category, etc., and another column for remarks may also be added.
- Numbering is not consistent as this is only a sample.

Second Report: Typical Equipment Manpower Project Summary Report

Form No, _____

Company Name & Address

Equipment Manpower Project Report

Equipment Project Summary Report

Section 1.1 Equipment Project Summary Report   Form No. _____

Location: _____   Month: _____

Classification/Position	Total Quantity	Day Shift	Night Shift	Remark
Manager (as maybe the case)				
Superintendent	1	1		
Supervisor	2	1	1	
Senior Engineer	1	1		
Engineer II, I (as the case maybe)	1		1	
Junior Engineer	2	1	1	
Lead Man	2	1	1	
Mechanic I,II	4	3	1	
Electrician/Electronic Tech I, II, III	3	2	1	
Machinist/Painter II	2	2		
Driver I, II, III	12	8	4	
Wheel Loader Operator I, II, III	6	4	2	
Tractor Operator I, II, III	5	3	2	
Others as the case maybe	xx	xx	xx	
Total	41	27	14	

Usually a more detailed report shall be prepared by personnel department for the monthly salary and wages. This personnel list shall come from the daily equipment time report or time card of each personnel and can be used to do the first and the summary reports above. To complete the report, each name shall mention the assigned equipment to them that month. This list should be used by the equipment project head to monitor those who may be transferred to other project, assigned other equipment, lay-off, etc.

3.4 EQUIPMENT MANPOWER CHART FOR TWO MAINTENANCE LEVELS (SEE PART III)

The equipment manpower charts point out how many personnel and manpower shall be for fifth and third echelon maintenance. The choice is dependent on the location of the project or how far the project is from the central shop. The numbers in the charts are dependent on the total number of equipment of the company assigned in the project.

The equipment head shall review the manpower list and adjust the actual manpower in the projects at least every three months. There should be immediate change if major movements take place and equipment population go beyond or below 20 units (Tables 3.4 and 3.5).

TABLE 3.4

Equipment Central Shop Manpower Chart for Fifth Echelon Maintenance Set-Up

Craft	20 Units		40 Units		60 Units		80 Units		100 Units	
	1st Shift	2nd Shift	1st shift	2nd shift	1st shift	2nd shift	1st shift	2nd shift	1st shift	2nd shift
Superintendent	1		1		1		1		1	
Clerk		1	1		1	1	1	1	1	1
Operations Section										
Truck Master	1	1	1	1	1	1	1	1	1	1
Operators	20	20	40	40	60	60	80	80	100	100
Drivers										
Tenders										
Maintenance Section										
Supervisor	1		1	1	1	1	2	1	2	1
Corrective Maintenance										
Lead Mechanic	1	1	2	1	2	1	2	1	2	1
Mechanic III	3	1	5	1	6	2	6	2	7	2
Mechanic 1	3	1	5	1	5	2	6	2	7	2
Lead Electrician	1	1	1	1	1	1	1	1	1	2
Electrician III		1	1	1	2	1	1	1	2	2
Electrician I	1	1	1	1	1	1	1	2	2	2
Electronic Tech I/II	1		1		1		2		2	
Lead Welder	1		1		1		1		1	
Welder III		1	1	1	1	1	1	1	2	2
Welder I	1	1	1	1	1	1	1	1	2	1
Lead Tire Man	1		1		1		1		2	1
Tire Man I/II	1	1	1	1	1	2	1	2	2	2
Machinist	1		1		1		1		2	
Preventive Maintenance										
Engineer I/II	1		1		1		1		1	
Jr. Engineer	1		1		2		2		2	
Clerk I/II	1		1		1		1		1	
Sr. Mechanic	1		1		2		2		3	
Mechanic I/II	1		1		1		2		2	
Sr. Electrician	1		1		1		1		1	
Lead Oiler	1		1		1		1		1	
Oilers	3	1	6	1	9	2	12	2	15	3
Warehouse Section										
Supervisor	1		1		1		1	1	1	1
Warehouse Clerk		1		1	1	1	1	1	1	1
Tool Keeper	1		1		2		2	1	2	1
POL Clerk	1	1	1	1	2	1	2	1	2	1

TABLE 3.5

Equipment Central Shop Manpower Chart for Third Echelon Maintenance Set-Up

Craft	20 Units 1st Shift	20 Units 2nd shift	40 Units 1st shift	40 Units 2nd shift	60 Units 1st shift	60 Units 2nd shift	80 Units 1st shift	80 Units 2nd shift	100 Units 1st shift	100 Units 2nd shift
Superintendent	1		1		1		1		1	
Clerk	1		1	1	1	1	1	1	1	1
Operations Section										
Truck Master	1	1	1	1	2	1	2	1	2	1
Operators	20	20	40	40	60	60	80	80	100	100
Drivers										
Tenders										
Maintenance Section										
Supervisor	1		1	1	1	1	1	1	1	1
Corrective Maintenance										
Lead Mechanic	1		1	1	1	1	2	1	2	1
Mechanic III	2	1	3	1	3	1	4	1	5	2
Mechanic 1	2	1	3	1	4	1	4	1	5	2
Lead Electrician	1		1		1	1	1	1	1	1
Electrician III			1	1	1	1	2	1	3	1
Electrician I	1	1	1	1	1	2	2	1	3	1
Electronic Tech I	1		1		1		2		2	
Lead Welder			1		1		1		1	
Welder III	1	1	1	1	1	1	1	1	1	1
Welder I			1		1	1	2	1	2	1
Lead Tire Man	1	1	1	1	1	1	1	1	1	1
Tire Man I/II			1		1	1	2	1	2	2
Machinist	1		1		1		2		2	
Preventive Maintenance										
Engineer I/II	1		1		1		1		1	
Jr. Engineer	1		1		2		2		2	
Clerk I/II	1		1		1		1		1	
Sr. Mechanic	1		1		2		2		3	
Mechanic I/II	1		1		1		2		2	
Sr. Electrician	1		1		1		1		1	
Lead Oiler	1		1		1		1		1	
Oilers	3	1	6	1	9	2	12	2	15	3
Warehouse Section										
Supervisor	1		1		1		1	1	1	1
Warehouse Clerk		1		1	1	1	1	1	1	1
Tool Keeper	1		1		1	1	1	1	1	1
POL Clerk	1		1		1	1	2	1	2	1

Part II

Equipment Administration Procedures

Equipment Administration
Procedure

4 Ownership Documentation

Documentation of equipment ownership is one of the most important works of the equipment department. With proper documentation, the department can have high flexibility on the equipment ownership from its acquisition to its disposal.

Decisions are faster and precise, given all documents in shipping, insurance, registration, usage, repairs, condition and the time and cost of each unit. The 201 folder will be the instance of the supplying documents or supporting references or records of the unit while in custody of the company.

Given a large number of equipment, the company can easily make demanding decisions when the time come to transfer or dispose of a unit at the end of the project. It will give the condition as well as the value of the unit at the end of a project it has been from and the final condition upon reaching full depreciation period.

To have a good 201 folder of equipment, the company must prepare the database work diligently and follow the order of documentation outlined in this manual. All are time and use tested. It shall give the management easy reference of record without losing time and opportunity.

4.1 REGISTRATION AND COMPANY CODE NUMBERS

Registration and company code numbering of equipment are the first mandatory documents prepared by the company. This entails the assignment of company code numbers to each purchased, delivered and received equipment. The code number sequence must be maintained regardless of the project location an equipment may have been delivered.

The Technical Services shall foresee to this implementation. Sequence of activities shall be company code number assignment and labeling or marking the equipment, registration, insurance, when needed, start a 201 folder and enter the equipment name and number to the company equipment database.

The company equipment number is the key to classify each equipment by a unique category and number within the types and classes of on and off highway equipment. Table 4.1 is an outline of equipment code system. For expediency, the equipment shall be divided into two main groups, major and support or minor equipment. Accessories such a bucket shall be made to be part of support equipment.

DOI: 10.1201/9781003360667-6

TABLE 4.1

Typical Classification of Equipment by Assignment of Company Code Numbers

Major Equipment

Property Codes	Classification	Property Codes	Classification
AC	Air Compressor	PR	Pneumatic Roller
AP	Aggregate Plant	SR	Smooth Roller
BE	Bituminous Equipment	SV	Service Vehicle
CE	Crawler Excavator	SVT	Service Truck
CF	Concrete Finisher	SW	Service Wrecker
CL	Crawler Loader	TE	Truck Excavator
CT	Crawler Tractor (Dozer)	TrBC	Bulk Cement Trailer
CP	Concrete Plant	TrDT	Bottom Dump Trailer
DE	Drilling Equipment	TrD	Dolly Trailer
DT	Dump Truck	TrHB	Hi Bed Trailer
FR	Foot Roller	TrLB	Low Bed Trailer
GP	Generating Plant	TrT	Tanker Trailer
LP	Light Plant	TM	Trenching Machine
MG	Motorized Grader	TT	Truck Tractor
MS	Motorized Scraper	WT	Water Truck
MT	Truck Mixer	VR	Vibratory Roller
PD	Pile Driver	WL	Wheel Loader
		WT	Water Truck

Minor or Support Equipment

Property Codes	Classification	Property Codes	Classification
AT	Air Tool	MSE	Machine Shop Eqpt
Bc	Buckets	PB	Power Broom
Ca	Concreting Accessories	JP	Jack Power
CM	Concrete Mixer	PS	Power Saw
Cv	Conveyor	CP	Compactor Plate
COP	Grout/Concrete Pump	CV	Concrete Vibrator
CS	Concrete Saw	SE	Shop Equipment
CS	Cement Silo	ST	Special Tools
CTa	Crawler Tractor Attachment	FL	Forklift
DE	Diving Equipment	TrV	Van Trailer
ET	Erectors Tools	WM	Welding Machine
LU	Lubricating Unit	WP	Water Pump

(Continued)

TABLE 4.1 (*Continued*)
Typical Classification of Equipment by Assignment of Company Code Numbers

Subclassification of Bucket and Crawler Tractor Attachment (Sample List)

Bucket		Dozer Attachment	
Clamshell	Backhoe	Rear Ripper	
Dragline	Shovel	Multi Shank Ripper	
Excavator	Grapples	Winches	
Material Handler		Pipelayer	
		Fork	

4.2 INSURANCE

Insurance is a mandatory requirement for on-highway equipment even if it may only operate inside the project restricted and controlled area. This is to ensure compliance to the country land transportation rules and regulations.

Wikipedia defines insurance as the equitable transfer of risk of a loss from one entity (the company) to another (insurance firm) in exchange for payment. It is a form of risk management primarily used to hedge against the risk of a contingent (project stringent condition or natural occurrence), uncertain loss. It is therefore prudent, especially if the project is in an uncharted territory. Still, this is the discretion of the company to insure any or all equipment under its fleet. On-highway equipment is however mandatory to be insured in almost all countries as part of the road safety regulation.

Insurance documents shall be part of the equipment 201 folder and database. If the policy is for finance department to keep the original documents, then a photocopy of it must be retained and filed in the 201 folder.

4.3 EQUIPMENT MODIFICATION AND ALTERATION

There are instances when some equipment will be modified or altered, changing its feature for another application. Example of such is a dump truck transformed to a water truck or motor scraper to water wagon, usually done for large projects such as dam construction. In this case, the equipment shall be assigned a new equipment code and number akin to its new form.

Job Order that authorizes modification or alteration as approved by the department head shall follow the drawings and specifications prepared and approved for this procedure. The process may be done at the central shop or by an outside authorized contractor. If an outside contractor be asked to do the process, the company procurement policy should apply.

5 Monitoring and Control

5.1 EQUIPMENT RECEIVING LIST (ERL)

All equipment of the company shall have an ERL regardless if the equipment is purchased new or used. This form shall be used for equipment received at the equipment central office or, in some instance, at the project site. Documents for bookkeeping shall be forwarded to accounting for company's fixed asset posting. All other documents such as warranty and guarantee papers shall be part of the equipment 201 folder.

Others such as operating and maintenance manuals and parts handbook, assembly instructions shall be given to the library. An inspection check list report prepared and declared as passing requirements by the equipment inspector must be submitted prior to official acceptance of the equipment.

All international shipments must be accompanied by the required documentation issued by proper governmental authorities. This includes, but is not limited to proper invoices, delivery receipts, custom clearance, with all obligatory government statements, measures and labels, packing lists which identify each carton and all other individual commodity, parts or manuals statements as may be required. A copy of this must be filed and be part of the 201 folder. Most important to check and ensure, there are proper warranty and guarantee certificates.

A copy of ERL is given to accounting for the asset register and bookkeeping purposes.

DOI: 10.1201/9781003360667-7

An example of ERL is given below.

Company Name & Address		
Equipment Management Department		
Equipment Receiving List (ERL) No. _____		
Date: _____		
Equipment Type:	Supplier:	P.O. Number
Equipment Number:	Delivery Date:	Invoice Number
Make	Serial No.	Engine Serial No.
Model	Chassis No.	Engine Year Model:
Documents: Bill of Lading O Delivery Receipt O Packing List O Air Way Bill O Custom Documents O Warranty Certificate O		
Description		
Accessories, Appurtenance and Components with the shipment:		
Inspected By: _____ Date: Received By: _____ Date:	Noted By: _____ Date: Approved By: _____ Date:	

5.2   EQUIPMENT DELIVERY AND SHIPPING LIST (EDSL)

EDSL is an internal document for the transfer of equipment from the central yard to the project or from project to project. It is confirmation of the transfer of equipment from the central yard to the project for assignment; also, it serves as the confirmation of the project receipt of the equipment. This shall also be used by projects when shipping back equipment to central yard or to another project for new assignment and deployment.

It contains the technical information, the list of attachments and accessories as maybe needed by the project. It will be accomplished every time equipment is transferred from the yard to a project or project to project. A copy of the document is sent back to the sender with the confirmation signature of the receiving end as proof of having received the equipment. There can only be one unit of equipment per EDSL.

Company Name & Address		
Equipment Management Department		
Equipment Delivery and Shipping List		No. _____
		Date: _____
Equipment Type:	Delivery Date	Project:
Equipment Number:	Receiving Date	Requested by:
Make	Serial No.	Insurance :
Model	Chassis No.	Conduction:
Equipment Status:(give detail if on breakdown)		
Description		
Accessories, Appurtenance and Components with the delivery:		
Documents with delivery: _____ Equipment 201 Folder _____ Parts Book _____ Job Order _____ Time and Cost File _____ Tire Service card _____ Accident Report _____ Registration _____ Inspection Check List _____ Oper. Manual Doc.		
Inspected By:_____ Date: Received By: _____ Date:	Noted By: _____ Date: Approved (Project/CSY) By: _____ Date:	

5.3   DAILY EQUIPMENT TIME RECORD (DETR)

The DETR card is the record of the equipment movement and the work performed by the equipment or on the equipment. The card shall be accomplished everyday every time equipment is on assignment at the project. Prepare one card each shift to make a distinction and record of the time during each shift, when there are two shifts. The data on this card shall be used for the time and cost record which in turn shall be used to determine the availability and utilization of the equipment. An equipment engineer shall be appointed in each project, depending on the number of equipment in the project (see manpower chart).

Company Name & Address Form No.

Equipment Management Department

Daily Equipment Time Record (DETR)

Project	Equipment No.	1st Shift	
		2nd Shift	
Location	Description	Time Start	
		Time end	
Date:	Start Meter Reading:		
	Finish Meter Reading:		

Time	Available		Breakdown			Work
	OPTG	IDLE	OR	NOR	PM	Performed

| Operator Name: | Equipment Checker: |
| ID No: | Site Confirmation: |

5.4 TRANSPORT TRIP TICKET

The trip ticket is used for the authorization to dispatch truck tractors, service truck vehicle and service vehicle to move personnel, materials or equipment from the central shop to any project or between projects. The ticket is used on daily assignment of the vehicle. It may be that a vehicle will take more than one day to travel and in which case the travel period from origin to destination must be indicated prior to authorization.

A ticket may have gone to more than one project in a single day hence can have more than one trip in a single trip ticket. In case of long trip, the date of arrival back at the place of origin will be the final period of the ticket.

The Transport Section that is responsible to secure the ticket for each trip daily must keep a log book to record each trip daily. The head of the section shall report to the central shop and yard for approval and administration. A file of completed trip tickets shall be kept at the Transport Office. The yard guard has to record the in and out of every equipment with their own log complete with log numbers of each trip ticket.

Company Name and Address									
Equipment Management Department									
Equipment No			Trip Ticket				Log No.		
Description:	Trip Description:						Date of Departure		
							Date of Arrival		
Trip No.	From: Origin	To: Destination	Time		Odometer		Driver Name	Purpose	Confirmation
			Depart	Arrive	Depart	Arrive			
Authorization By: Name: Designation:									

5.5   EQUIPMENT TEMPORARY TRANSFER ORDER

This is the document and control for the temporary transfer of equipment from project to central yard or from one project to another and is used for monitoring the temporary equipment movement from its original assignment. Approval of transfer must be secured from the central yard prior to the transfer. If projects are close to each other, it will be prudent, if the usage of one will be longer than one month, to ensure coverage of the transfer or temporary use of the other project to be reflected in the equipment monthly report.

Instances happen when an equipment has been transferred by project A to project B without document and if the person who authorized the transfer at A leaves the project, the inventory list becomes a confusion of equipment listed in two projects or missing in one project, or worst missing in both, one physically the other documentary wise.

Company Name & Address		Form No.		
Equipment Temporary Transfer Order				

From : _____   To: _____   Date:_____

Location: _____   Location:_____

Equipment below is authorized to be transferred for a period of _____

per attached request by _____.

Equipment No.	Description	Make	Model	Serial No
Remarks:				

Approved by   Received by:

5.6   MISSING/DAMAGE ACCESSORY REPORT

The report shall state any missing or damage part of equipment or its return to the central yard or to the project storage or parking area. If equipment is returned in the central yard and upon inspection was found to be with missing assessor, a missing report shall be attached to the inspection report for proper disposition. A copy of the report shall be given to the returning project for explanation and to accounting department for assessment and proper charging of loss.

Company Name/Address			Form No.	
Missing Damage Accessory Report				
Equipment No.		Description	Date:	

Above equipment was received at the central yard per EDSL No. _____ dated _____

With the following unstated and missing/damage accessories:

No.	Description & Part No.	Unit	Quantity	Remarks

Please advise reason and where about of accessories above.

None advise in 2 days shall prompt the charging the project with full replacement cost.

Inspection Report Attached	
Prepared by:	Approved by:
Noted by:	

6 Equipment Standard Operating Procedure

6.1 STANDARD OPERATING

This procedure must be an essential part of each operator/driver start-up route to ensure maintaining the equipment good condition. Therefore, they must be strictly observed and followed by all drivers, operators and tenders before, during and after operation of their respective assigned units. The process must be part of the testing of drivers and operators.

6.1.1 BEFORE STARTING THE ENGINE

1.1 Walk around the equipment or vehicle to inspect for loose or missing parts, cracked or damaged components, fluid and lubricant leakage, abnormal wear, tire inflation, etc.

1.2 Check all oil levels, water level, electrolyte, fuel level, radiator cooling fins, air cleaner, etc.

6.1.2 ON STARTING WARMING UP THE ENGINE

2.1 Strictly observe and follow standard procedures outlined in the operator's manual. As a rule of thumb, engine must be idled three to five minutes before it is put to operation.

2.2 Upon starting the engine, immediately observe all gauges for proper functioning.

2.3 When all gauges indicate normal readings, check brakes and steering wheel. Operate hydraulic system for two cycles to ensure proper functioning.

2.4 In case where one or more gauges indicate abnormal reading, shut off engine and immediately notify the Preventive Maintenance Crew.

2.5 Similarly, report to the Preventive Maintenance Crew that unusual noise coming from the engine or power train.

Procedures 6.1.1 and 6.1.2 above should be conducted within 20 minutes before the start of shift.

DOI: 10.1201/9781003360667-8

6.1.3 DURING OPERATION OF THE EQUIPMENT

3.1 Observe safety precautionary measures at all times.

3.2 Be alert for all unusual noises emanating from the machine and frequently check all gauges.

3.3 Never pit one machine against another for each has its own weakness and limitations to avoid unnecessary wear and undue breakdown.

3.4 Study and master the controls and operations of the machine in order to get the right feel for maximum efficiency.

3.5 In case of failure while in transit (for vehicle only), report the incident to the nearest Project Equipment Superintendent and, if possible, inform the Divisions Manager.

6.1.4 AFTER OPERATION OF THE EQUIPMENT (ENGINE SHUTOFF)

4.1 Park the machine on a safe level area.

4.2 Set all safety devices and correctly position all levers.

4.3 Run the engine at idle three to five minutes.

4.4 Carry out walk around inspection (repeat 1.1).

4.5 Report promptly any suspected defect or defects of the machine to the Preventive Maintenance Crew.

6.1.5 DURING PREVENTIVE MAINTENANCE SERVICES

5.1 Oversee the proper servicing of the machine and, if necessary, assist the PM Crew in locating vital lube points.

5.2 Make sure that cleanliness is observed by the PM crew at all times.

5.3 Check oil, fuel, electrolyte, and coolant housing levels.

5.4 Inform the servicing mechanics of noted defects of the machine.

6.1.6 DURING MACHINE BREAKDOWN

6.1 Assist the mechanics in the repair of the machine you are held accountable for.

6.2 Complain other defects of the machine for corrective measure.

6.1.7 DURING MACHINE SHUTDOWN CAUSED BY UNFAVORABLE WEATHER AND WORKING CONDITION

7.1 If the situation warrants, conduct body an under-chassis cleaning of the machine.

7.2 Carry out some corrective check-ups. However, all are subject to the central yard policy.

6.2 EQUIPMENT ROAD LIMITS AND TRAFFIC RULES

To conform to the highway and road rules of the country, management must know the weight and size of each equipment especially of on-highway equipment such as but not limited to service vehicles, truck tractors and trailers to ensure conformity with the road limit law of the country. Should equipment be moved from one country to another, the road rules of both countries must be readily available to management for immediate compliance.

Examples of such rules are the permissible weight and dimensions of vehicles in highway traffic and special fees for the use of highways. Each country has its own rules and must be checked thoroughly.

6.2.1 TYPICAL PERMISSIBLE MAXIMUM WEIGHTS

1. Per most heavily loaded wheel – 3,600 kg.
2. Per most heavily loaded axle group – 14,500 kg.

(Note: Two axles of the group being at least one meter and less than two meters apart.)

An axle weight shall be the total weight transmitted to the road by all wheels, the center of which can be included between parallel transverse planes on meter apart extending across the full width of the vehicles.

6.2.2 TYPICAL PERMISSIBLE DIMENSIONS

1. No vehicle operating as a single unit shall exceed the following dimensions:
   a. Overall Width   2.5 meters
   b. Overall Height   4.0 meters
   c. Overall Length:
      i.   Freight vehicle with two axles   10.0 meters
      ii. Passenger vehicle with two axles   11.0 meters
      iii. Vehicles with three or more axles   14.0 meters
2. No vehicles and/or trailer combination shall exceed 18.0 meters in overall projected length, including any load carried on such vehicle and trailer.
3. No articulated vehicles shall be allowed to draw or pull a trailer, and no vehicle already drawing a trailer shall draw another.

6.2.3 TYPICAL SPECIAL PERMITS FOR VEHICLES TO OPERATE ON PUBLIC HIGHWAYS

1. To operate any vehicle or trailer outfit with axle, wheel or axle group loads in excess of the limits in (1) above, or per exclusive country limits.
2. To operate a vehicle the size of which exceeds the limits of permissible dimensions above.

3. To operate a motor vehicle with any part of the load extending beyond the projected width of the vehicle.
4. To pull two trailers behind a vehicle.
5. For any other special authority relating to the use of vehicles, not otherwise specified and provided above.

6.2.4   TYPICAL PERMITS FOR MOVEMENT OF OFF-HIGHWAY VEHICLE ON HIGHWAY USE

Prior movement of off-highway vehicle on highways traffic, conduction permit shall be first obtained from the transportation/traffic office of the country.

This can be obtained by the liaison of the Registration Section of the department.

6.3   EQUIPMENT LIMITATIONS AND CARGO LOADING

Every country has its own truck, cargo and road load limitation regulations. Equipment department must see to it that it has copies of all these and be guided accordingly for an effective and efficient movement of equipment from one location to another location.

6.3.1   SIGNIFICANCE OF LOAD DISTRIBUTION

Distribution of cargo has a very definite bearing on the life of tires, axles, frame and other parts. The fact that the truck or trailer is not loaded beyond if its gross weight capacity does not necessarily mean that the individual tires or axles may not be overloaded by faulty distribution of cargo. Faulty load distribution adversely affects safety, cost of operation and life expectancy of freight vehicles.

6.3.2   METHODS OF DETERMINING CENTER OF LOAD

As an aid to properly loading a truck or semi-trailer units, the center of payload must be determined. In a truck, the center of payload is the center of the body or the midway between the rear of the driver's cab and tailgate. In a truck tractor semi-trailer unit, the center of payload is roughly the center of semi-trailer body because the tractor's front wheels seldom carry any of the payloads.

When loading, it is important that the maximum capacity of the vehicle not be exceeded over any one of the axles and, if possible, that loads are distributed so that there is less-than-maximum axle loading. This principle may not apply in hauling of equipment and materials on low-bed trailers where the load generally is placed toward the rest of the trailer due to the greater number of axles and tires available on the low-bed trailer for weight distribution

6.3.3   MAXIMUM PERMISSIBLE PAYLOAD TRUCKS

The trend of automotive industry is to rate trucks in terms of maximum gross vehicle weight rather than by tonnage capacities. This has resulted in poor utilization of trucks because of loading below rated capacities and confusion in many instances as to the maximum pounds of payload that may be transported in trucks rated in tonnage capacities.

6.3.4 DETERMINATION OF MAXIMUM PAYLOAD

The maximum payload permissible is determined by deducting the curb weight and the weight of the driver (170 lbs.) from the manufacturer's gross vehicle rating. The maximum gross weight rating for a specified operating condition applies only when the tires and equipment on the truck are in accordance with, or exceed, the manufacturer's recommendation for the specified operating condition, referred to hereafter as "ideal" or "severe".

6.3.4.1 Recommended Payload under Varied Operating Conditions

Some trucks may be utilized at projects at which the maximum gross vehicle weight rating does not apply because the operating conditions may be more severe than those for which the gross vehicle weight rating was established. In such cases, the percentage shown under this chapter is the payload limitations recommended for each operating conditions:

6.3.5 IDEAL CONDITIONS

Ideal conditions mean that the truck is operating over improved level roads, such as asphalt or concrete, at constant relatively moderate speed, with no adverse weather or road conditions. Under these conditions, recommended payload equals *100%* of the maximum permissible payload.

This requirement may differ from country to country, but it is prudent to check and inquire rules and regulation to this effect. Some countries have manuals and guide book outlining these regulations and rules for reference. Copies in this case must be secured and be readily available for reference.

6.4 MARKING OF EQUIPMENT CLASSIFICATION AND NUMBER

Company equipment shall all be marked with a logo and the equipment classification and the number it has been assigned. The logo and equipment number must be painted or sprayed at the driver cab's left and right sides of the equipment. On-highway equipment shall further be marked with its registered number at the rear left and rear right sides of the equipment body (Table 6.1).

TABLE 6.1
List of company equipment to be controlled with assigned numbers

All Service Vehicles	All Tractor Trucks
All Service Trucks	All Heavy Equipment
All Dump Trucks (on-highway type)	All Minor Equipment
All Mixer Trucks	

Further to this, the registered passenger, gross weight and net weight capacities shall be marked plainly and conspicuously on both sides thereof, in letters and numerals not less than five centimeters in height.

Passenger and weight capacities could be made available upon request to mark all equipment assigned for central yard operation.

6.5   PROPER USE OF COMPANY VEHICLES

To ensure good housekeeping and good conduct, there must be rules on the proper use of company vehicles which must be followed and complied by all company personnel:

1. No riders are allowed on all Service Vehicles, Service Trucks, Dump Trucks, etc., unless authorized by the Truck masters.
2. Avoid overloading specially on service pick-ups.
3. Exercise proper discretion and behavior while driving.
4. Company vehicles shall not be used in any case other than official business.
5. All company vehicles shall be covered by trip tickets stating purpose, user, destination, etc.
6. Before operation, drivers shall systematically inspect their vehicles on the condition of the battery, tires, fan belts, radiator water level, engine oil level, panel gauges, horn, lights, etc.
7. Driver should assist maintenance personnel in the proper servicing and maintenance of vehicles.
8. Drivers should know and obey country traffic signs, rules and regulations.
9. Drivers should ensure that vehicles are properly kept and clean at all times.
10. Only official drivers are allowed to operate company equipment.

6.6   CONTROL OF VEHICLE EMERGENCY
TOOLS AND ACCESSORIES

Emergency tools are mandatory part of any vehicle. It shall be provided upon purchase of the equipment and must be retained with the equipment regardless of whoever it has been assigned to. It shall be part of the turn over when equipment is handed over from one operator/driver to another. For better control, they must be physically traceable from records that shall be responsible for their proper use and safe custody.

The emergency tools covered under this are as follows:

1. Jack (indicate the capacity according to type of equipment).
2. Tire Wrench.
3. Tool kit included during receipt of new vehicles (i.e., adjustable wrench, open wrench set, pliers, and screw driver).

4. Early Warning Device.
5. Other equipment implemented as indicated in the equipment delivery and shipping list (EDSL).

In projects that have no designated Truck masters, implementation of responsibility is delegated to the Equipment Supervisor or others as designated by the Projects Maintenance Heads. This is important especially if the project is composed of different sites and is far from the central or project shop.

Lost and damaged tools should be made known immediately to the Truck master who shall conduct investigation, if necessary, of the actual cause.

6.6.1 CONTROL OF EQUIPMENT ACCESSORIES

For road safety, equipment accessories and parts must be inspected and accounted for each time there is a transfer of assign operator/driver. Missing and/or damaged accessories must be minimized if not totally eliminated.

For better control, accessories must be physically available and contained in the EDSL.

The equipment accessories covered are the standard accessories of units and engines but that not limited to are as follows:

- Unit accessories – side view mirror, rear view mirror, windshield wiper and motor and signal light switch.
- Engine accessories – starter, generator, alternator and voltage regulator. GPS when declared available.

Control shall be as follows:

1. Drivers/operators who shall accompany the shipment shall check the physical availability of the accessories listed in the EDSL.
2. Drivers/operators who shall accompany the shipment shall be made to sign the EDSL to certify that accessories listed therein are physically present.

6.6.2 EQUIPMENT INCOMING CONTROL

Equipment shall be physically inspected against the incoming EDSL and the outgoing EDSL. Accessories listed in the outgoing EDSL but not included in the incoming equipment shall be considered missing.

Drivers/operators who accompanied the shipment shall be present during the inspection and shall only be allowed to return to the project on completion of the inspection.

In case of any discrepancy, a Missing/Damaged Accessory Report shall be prepared and project concerned shall be advised immediately. Report shall be attached to the project's copy of EDSL.

6.6.3 PROJECTS

6.6.3.1 Equipment Incoming

a. Equipment shall be physically inspected against the Incoming EDSL, and accessories listed therein not physically available shall be considered missing.

b. Drivers/operators who accompany the shipment shall be present during the inspection and shall only be allowed to return to central yard or project on completion of the inspection.

c. Drivers/operators shall sign the EDSL after inspection.

d. In case of any discrepancy, a Missing/Damaged Accessory Report shall be prepared and project concerned (or central yard) shall be advised immediately. Report shall be attached to the sending project's copy of EDSL.

6.6.3.2 Equipment Outgoing

a. List down all equipment accessories on the EDSL with complete description such as quantity, make, model and serial number. The incoming EDSL shall be made as reference in listing down the accessories. Appurtenances and other detached or detachable accessories shipped or left behind shall also be indicated in the EDSL.

b. Drivers/operators who shall accompany the shipment shall check the physical availability of the accessories listed in the EDSL.

c. Drivers/operators who shall accompany the shipment shall be made to sign the EDSL to certify that accessories listed therein are physically present.

- Lost   • 100% of cost of exact replacement
- Damage   • 50% of cost

Part III

Equipment Maintenance

7 Equipment Maintenance System

Maintenance services cover almost all facets of equipment movement prior, during and after its field operation. It starts in the efficient accomplishment of all inspection, preventive maintenance (PM), repair, overhaul and reclamation necessary to ensure the equipment is in good condition to meet operating requirements again and again.

It is apparent that this objective is primarily an operating or doing function and that reasonable accomplishment of certain aspects of maintenance's objective will require instructions and directions from a planning and control function.

To follow a consistent and tested PM schedule ensured the equipment performance can be tracked and monitored dependably in an orderly manner. No sudden disruptions, delays or surprise failures. This is a long-term advantage.

As a result two most important elements, availability and cost are always in focus without much inconvenience. Availability will be a guide to equipment accessibility and operability for the field operation, while the cost will be the reduction of unscheduled breakdown and delay or stop operations. These can be from early detections of parts fatigue and failure, such as filters, seals, fittings, belts, and quality of oil, water, and air. Small as they may seem, can bring the equipment to a stop. A reliable and consistent PM has been proven can reduce repair cost by 25% on maintenance alone.

Longer equipment economic life- When equipment is being checked and maintained, there will be no surprises, especially in the middle of operations. The equipment is kept in its best conditions always. With routine preventive maintenance, the equipment performance and, importantly, equipment life are extended

The service also includes detailed understanding of petroleum, oil and lubrication (POL) and their undercarriage services, two common high value items of equipment. The understanding of these items can bring much order and cost-efficient operation.

7.1 FUNCTION OF MAINTENANCE

The basic function of maintenance can be summarized as the accomplishment of all upkeep work indispensable to keep and maintain equipment in a condition to ensure its availability to be utilized by the project as per schedule. This broad function can be grouped into subdivisions of (1) inspection, (2) preventive maintenance and (3) corrective maintenance which comprise of repair, overhaul and reclamation.

DOI: 10.1201/9781003360667-10

7.1.1 INSPECTION

Inspection is the first maintenance equipment received from the moment equipment is delivered to the company as part of its acquisition. Upon acceptance, the equipment shall have its periodic inspection to ensure safe efficient operation. Inspection shall be performed every time equipment shall be sent out of the project and/or received by another project. Proper documents must be accomplished to indicate the status of the equipment.

Daily inspection prior to operating the equipment shall also be performed by the driver/operator as a first step to operating the equipment. This is mandatory and compulsory. This is also making certain that equipment is operation-ready daily. It shall be the responsibility of the driver/operator to report any defect or malfunction of the equipment to his superior.

In addition, inspection has the responsibility of making certain that all equipment requiring periodic maintenance or overhaul receive attention and are checked at the required time. Maintenance inspection has the clearly defined responsibility to check the recorded utilization of equipment and the issuance of orders to withdraw each from service for maintenance repair at the schedule time.

7.1.2 PREVENTIVE MAINTENANCE (PM)

Properly defined, true maintenance work is confined to the checking, adjustment, routine replacement, lubrications and clean-up necessary to make certain the equipment are in proper condition and ready for use. This maintenance work is predictable, readily adaptable to accurate planning and scheduling and can be placed in a standard time basis for cost-control purposes. Preventive maintenance is the correct way to pre-diagnose possible and inevitable equipment failure before they occur or develop into major defects.

It is largely classified as preventive maintenance and should not be confused with the unpredictable work load generated by breakdown and trouble calls. It is the task of preventive maintenance, through planned, scheduled inspection, and overhaul to ensure that equipment failure should be anticipated and prevented. The planning is primarily based on the equipment manufacturer's tolerance established for each fitting, part, sub-assembly and assembly that makes the whole of the equipment.

Not all project conditions are the same. Some are located in the desert or arid condition, some in very humidity area. Fittings and filters may need shorter replacement period in one than in another. Maintenance engineers must know these conditions and its effect to their equipment. Other conditions that must be checked are the pH property of water used for the cooling system. This may bring corrosion and the need to be corrected immediately to prevent unscheduled failure. If this prevails, then corrective maintenance becomes inevitable.

7.1.2.1 Preventive Maintenance (PM) System

Preventive maintenance (PM) should not be confused with the unpredictable work load generated by breakdowns and trouble calls. PM is the area in which maintenance can affect the greatest savings in overall equipment operating costs.

Downtime, the period in which equipment is out of service hence not generating revenue, is a large factor in company cost. Equipment suddenly rendered unserviceable because of breakdown is not only costly in terms of lost production, but the total cost may be greater than the apparent loss because of the effect on project schedule. It is the task of preventive maintenance, through planned, scheduled inspection, maintenance and overhaul to ensure this does not happen.

PM is the periodic maintenance based on cumulative operating hours of the equipment to uncover condition leading to possible break down. This is to upkeep the equipment by timely replacing worn-out fittings and seals, adjust or repair such to tenacious condition according to the tolerance limits established by the equipment manufacturer.

PM is also to find out if oil and lubricants lose viscosity, fil ters have worn out filament, seals, bearings and bushings are thinning, contracting or reducing their protective shell.

7.1.2.2   Objectives of a Preventive Maintenance (PM)

Preventive maintenance is a scheduled deterrent work on the equipment to anticipate kits, fittings, parts and sub-assemblies breakdown due to its normal materials fatigue or weariness. PM is performed when oil and lubricants lose viscosity, filters have worn out filament, seals, bearings and bushings are thinning, contracting or reducing their protective shell.

To achieve this, efficient equipment maintenance program must be prescheduled and people, particularly engineers, on hand should be put to service the program. Here are major outcomes of a good PM program:

1. Fewer large-scale repairs, and fewer repetitive repairs, hence efficient use of manpower and facilities.
2. Lower repair costs for simple repairs made before breakdown because less manpower, fewer skills are needed for planned shutdowns than for unscheduled breakdown.
3. Increase productivity due to extended equipment life expectancy.
4. Attain 80% of equipment availability at all times.
5. Decline of maintenance cost – labor and material-on equipment in the program.
6. Identification of equipment with high maintenance costs, leading to investigation and correction of possible causes such as (a) misapplication, (b) operator abuse and (c) wrong technical support or materials such as high water pH.
7. Shift from inefficient "Breakdown" maintenance to less costly scheduled maintenance, hence effective work control.
8. Better industry relations because operators and drivers do not suffer involuntary layoff or loss of incentive bonus from breakdown.
9. Better scheduling of equipment thereby maximized project efficiency and accomplishments.

7.1.2.3   Contents and Scope of Preventive Maintenance

There are four basic contents and scope of preventive maintenance to consider in order to effectively carry out each Preventive Maintenance (PM) echelon, namely:

1. Inspection
2. Lubrication
3. Periodic adjustments
4. Replacements

7.1.2.3.1   Inspection

Equipment inspection is conducted to determine which parts are needing lubrication, adjustments, minor repairs, extensive repairs, replacements, etc. Inspections during PM operations shall use the inspection check list as given in Part IV. The necessary maintenance shall be done during PM execution as inspection so indicates. A final inspection shall be made before releasing the equipment to operation to check whether the noted deficiencies are properly attended.

Inspection responsibility is delegated to the Project PM engineer who shall in turn advice the PM crew of his findings. In any case, the mechanic and electrician of the PM crew shall assist the inspector during inspection.

Inspection shall be done in two fashions: 1) visual inspection when the equipment is at stand still and 2) load test analysis when the equipment is operating. The first type of inspection will determine the physical condition and appearances of the equipment while the second type will determine the mechanical condition of the equipment which could not be detected through visual inspection.

Operators and drivers are the first line of defense against equipment wear, failure and damage. Equipment must be inspected systematically before, during and after operation so that defects may be discovered before they result in serious damage or failure. Defects discovered during these inspections or during operation of the equipment will be noted at the back face of the Daily Equipment Time Record and reported as soon as operation has been stopped. The operator and drivers must stop operation immediately if a deficiency is observed that could damage the equipment or render it unsafe.

7.1.2.3.2   Lubrication

Lubrication should not be construed maintenance; it is one of the important elements of PM. However, PM to be more effective requires that lubricant is applied at the proper time, the right amount and the correct grade. Over- or under-lubrication is as detrimental as no lubrication at all. This will be discussed in detail later.

7.1.2.3.3   Adjustments

Periodic adjustments involve the scheduled tune up of engine, resetting of hydraulic valves and other assemblies and sub-assemblies recommended by manufacturer to be adjusted at periodic intervals. This simple procedure can save the company a lot of time and expense. Most equipment part must be airtight shut to maintain the system operating, and any loose or leaking system may fail due to lost pressure and air, gas or lubricant as the case maybe.

Tightening procedure has to be performed and checked consistently, ditto to maintaining tolerances of moving parts and accessories of the machine assembly of the equipment. Equipment manufacturers provide this in their operating and maintenance manuals and guidebooks.

7.1.2.3.4   Replacements

Parts replacements entail the replacement of defective and worn-out parts as inspection so indicate and those parts recommended by manufacturers to be replaced at specified intervals. Parts found to be defective and worn out during inspection shall be checked with warehouse for availability otherwise prepare a requisition. Replacement shall be effected on arrival of parts if in maintenance's opinion continued use would pose danger and safety hazard both to life and property. However, if continued use does not pose any danger or safety hazard, then replacement could be made the next subsequent PM schedule. A minimum level of parts inventory shall be maintained by Warehouse for PM.

7.1.2.4   Preventive Maintenance (PM) Program

The preventive maintenance program shall be implemented by the technical services department. Basically, the program of each and every unit of equipment is adopted from the recommendations of the equipment manufacturer. The initial preventive maintenance of equipment is practically drawn from the equipment manufacturer's recommendations.

The voluminous information and feedbacks collected indicate a need to deviate from the manufacturer's recommendation according to local project climatic conditions, equipment application and working conditions. The revised and modified PM program hereto would dramatically reduce the costs of maintenance, both labor and materials, and increase equipment productivity.

7.1.2.4.1   Preventive Maintenance (PM) Time Reference Schedule

7.1.2.4.1.1   Maintenance Schedule by Echelon System with Spare Parts and Assembly List (See Attachment 5)   Maintenance is taken to keep material or equipment in serviceable conditions to restore it to operating condition when it is unserviceable. It denotes all work functions are performed on vehicular and earthmoving equipment to determine, prevent and correct physical damage and mechanical malfunctioning in the degree necessary to continue or restore equipment to a safe and serviceable condition. The term "maintenance" includes the functions designated as inspection, repair, testing, service, rebuild, overhaul and reclamation, and classification as to serviceability (see Attachment 5).

To organize the maintenance of equipment from the time of acquisition to its operation from field work to field work, an echelon scheme has been developed which will be carried out for a progressive maintenance schedule according to the net utilization time of the equipment. The purpose is to ensure that the equipment is checked at certain time intervals for checking and testing of equipment parts and assemblies. In this way, any developing defects can be detected immediately before it brings failure to a part or assembly of the equipment.

As the echelon goes further, the inspection becomes more thorough and demanding. It begins with basic inspection and will end with engine and under-body assemblies overhauling sequence. Planning and scheduling of time data shall begin from the daily time record to the maintenance schedule and to the time and cost record of each equipment.

7.1.2.4.1.2   First (1st) Echelon Maintenance or PM 1   Minor repairs that can be accomplished by operators and driver using tools provided with the equipment consist of tightening of hardware, loose connections and cleaning, or in the case where moving mechanics are available repairs identical to the above and limited by the tools available with mechanics. This also includes the daily services done by the PM crew. A time limitation is usually set of a maximum of one (1) an hour for repair or repairs required. In all cases, the operator shall assist the roving PM mechanic.

7.1.2.4.1.3   Second (2nd) Echelon Maintenance or PM 2   It is usually main-tenance made in site which can be accomplished as in the last echelon but with addition of fitting readily replaceable external parts such as fan belts, covers, missing bolts and nuts. Checking water separator (if equipped), draining water for sediments and checking brake and clutch performance, hydraulic pressure and transmission for leaks and seepages are some items for this period. Maintenance of this nature should have a time limitations of two (2) an hour. In both 1st and 2nd echelon repairs, the operator will assist the roving mechanic or workshop personnel dispatched to the jobsite.

These repairs should be confined to operational and safety requirements only, and if they do not meet this requirement, third (3rd) echelon repairs must be considered, i.e. repair must satisfy requirements and not induce further damage.

7.1.2.4.1.4   Third (3rd) Echelon Maintenance or PM 3   This may be done in site but are usually and referable performed at the project workshop. They consist of maintenance of 1st and 2nd echelon nature plus the replacement of readily replaceable sub-assemblies or kit repair procedures. This also includes trouble-shooting and preventive maintenance. Some items to check are suspension and steering for parts replacement, drive chain for wear and parts replacement, engine tune up, oil drain and oil filter change. Again in a maximum time limit of not more than 5 hours in site or more than 25 hours in the workshop should be set. Exception to the extent of onsite repair should only be resorted to if the econom-ics so indicate, taking into considerations urgency or requirements, cost of trans-portation, work force required, tools requirements, etc. In other words, each one must be assessed individually.

7.1.2.4.1.5   Fourth (4th) Echelon Maintenance or PM 4   These are defined as the major overhaul of sub-assemblies. These may require the repair of replace-ment of internal parts. Some items for check are for engine top overhaul check and diagnosis facing and lining, hydraulic pressure oil and lubrication check and abnormal wear of clutch, noise or knock or oil leaks from rear axle, suspension system. This is usually but not necessarily a part of the 5th echelon workshop, i.e. one can visualize a sub-workshop devoted solely to the repair and recondition of certain types of certain types of assemblies. Time limits are not set for this type of repair except in general terms by use of flat timing or that gained by normal experience, if flat rate timing or that should be used initially and modified from experience under local project condition.

7.1.2.4.1.6 Fifth (5th) Echelon Maintenance or PM 5

This is the major repair and overhaul entity represented in the central shop. They are responsible for all repairs in excess of list and 3rd echelon. This includes the major repair and renovation of all equipment and their components. This type or variations of the type of repair may be done in 3rd echelon or field workshops. However, this decision must be based on numbers and skills of tradesmen required, effect on normal repair functions, availability of spare parts, estimated economy or savings involved, etc.

In some cases, it might be advisable and advantageous to provide additional staff and tooling from central shop for the duration of the repair. Most equipment manufacturers will give the vital number of total cumulative operating hours for overhaul procedure typically starting from 6,000 hours to 10,000 hours for major overhaul, depending on type of engine and working condition.

Suggestion for project work refers particularly to remote or projects far from central shop where transportation costs may be a major factor. Responsibility for various echelon maintenance such as 1st, 2nd and 3rd echelon maintenance are the responsibility of field or project workshop and staff. Some of the following types of repair may be considered as normal 3rd echelon responsibility subject to time limitations of 25 man hours.

Top overhaul, cross joint replacement, brake repair and re-lining, clutch replacement, etc. that require 4th and 5th echelon maintenance may be delegated to project workshops on individual or case basis. Repairs made by central shop shall include all 1st and 3rd echelon maintenance.

The application of the echelon system is PM scheduling according to daily, weekly, bi-weekly, monthly, bi-monthly and fourth monthly check-ups and servicing. For the purpose of course putting into consideration the lubrication and scheduled maintenance services recommended by the equipment manufacturers, the following schedule is projected.

Utilization Period	Operating Hours	PM Schedule
Daily	10 Hours	Inspection
Weekly	60 Hours	PM 1
Bi-weekly	125 Hours	PM 2
Monthly	250 Hours	PM 1–3
Bi-monthly	500 Hours	PM 1–4
Fourth monthly	1000 Hours	PM 1–5

In the true sense, the date scheduling by hour is actually only preliminary programming. The real scheduling takes place when a definite day has been set, and the job has been planned as to method, tools, materials, equipment and work location.

7.1.3   Equipment Corrective Maintenance

Corrective maintenance shall be performed after the maximum operating time and must be overhauled in accordance or following the Original Equipment Manufacturer (OEM) manual. Since the most permanent parts of an assembly are located in the internal section of the assembly of an engine or pump or transmission or axle, the assembly has to be overhauled to change the replaceable internal part not reached by PM, parts like bearing, shims, gaskets, compression ring and exhaust valves. The maintenance is to save the lives of the main components which are the principal parts of the assembly.

Corrective maintenance requirements are like PM, planned ahead, and are organized up to a selection of mechanics, electricians, parts, and assembly. The overhauling section of the shop can be structured to provide the most difficult engine repair and overhaul works.

Corrective maintenance if well scheduled, therefore anticipated, will not be cumbersome and a simple normal process that is basic maintenance based. The end results shall be no unscheduled failures and stoppages at any time during operations.

There are three stages of corrective maintenance, and each shall be part of the equipment maintenance schedule in the central or field equipment shop. The repair, overhaul and rehabilitation maintenance are subsequent repairs needed due to parts material failure which comes after long and constant period of utilization of the equipment. Equipment assemblies and parts have predicted length of utilization time after which the material shall show natural fatigue and then failure.

Manufacturers established these parts and assemblies measure and criteria as allowable tolerance which should not be exceeded to prevent major failure of the equipment. This tolerance rating goes for every fitting, part, sub-assembly and assembly just as it has been made reference for preventive maintenance.

7.1.3.1   The Three Types of Corrective Maintenance

First: Repair means to restore to a serviceable condition of such parts, sub-assemblies that can be accomplished without completely disassembling the unit. Corrective repair to alleviate unsatisfactorily conditions found during preventive maintenance inspection is considered a portion of this work. Repair is that unscheduled work, often of an emergency nature, necessary to correct breakdown, and it includes troubleshooting. These are maintenance functions which restore individual parts, components or assemblies of an equipment unit to a safe, serviceable condition. The term "repair" includes unit rebuild and unit replacement.

7.1.3.1.1 Repair as Necessary

These are repairs made to correct known faults or specific items reported by the project user. Normally, these repairs fall within the 1st and 3rd echelon responsibility provided time factor not exceed 3rd echelon limitations, i.e. 10 man hours. This type of repair only guarantees the specific item repaired will give satisfactorily service after minor restore of failed, usually external, parts.

Second: Overhaul or rebuild is considered as the planned, scheduled reconditioning of equipment. This work will always involve one or more of the elements to disassembling, tear down, examination of tolerances, replacements of parts or reconditioning, reassemble and testing to restore into a condition comparable to new.

All planned overhaul should be handled by a maintenance inspector who reviews the job and determine the complete bill of material. The part or material should then be ordered for delivery prior to the scheduled time of the overhaul. When the replacement parts are available the equipment to be overhauled should be withdrawn from service at the scheduled time for maintenance.

7.1.3.1.2 Overhaul as Necessary

It denotes all major repairs required to correct known faults and those found necessary by visual and operating inspection. Particular emphasis is placed on the aspects of safe operation. This type of repair again includes all lists and 3rd echelon should not be exceeded, i.e. 25 man hours. Per-inspection time and out inspection are not included.

No attempt is made to repair within manufacturer's tolerance except for safety devices. This type of repair provides reasonable assurance that the equipment will give satisfactorily performance.

7.1.3.1.3 Progressive Types of Overhaul

7.1.3.1.3.1 Top Overhaul This refers to engine and piston – type compressors only. It consists of repair replacement of cylinder heads and the repair or replacement of its components, i.e. valve gear, oil lines and gasket.

7.1.3.1.3.2 General Overhaul This type of repair is similar to but more extensive than repair as necessary. It ensures that all components of the equipment have been inspected and all necessary repairs have been completed to provide considerable further satisfactorily service. Some, but not necessarily all, components will be repaired to within the manufacturer's limits and tolerance. In particular, all safety aspects of the repairs made are within accepted tolerance.

7.1.3.1.3.3 Major Overhaul This type of repair will bring all components within the limits of the manufactured tolerance and will probably provide equivalent of the same type in so far as satisfactory, economical service is concerned. It should be noted that this type of repair is usually very costly and require most careful assessment of all the various factors before it is authorized. Some of these factors are believed to be estimated cost of repairs in relation to depreciated cost, replacement cost and future earning capacity.

Third: Rehabilitation is the process of restoring, repairing or fabricating salvaged or unserviceable articles or component parts. This is a major repair and may mean complete disassembling of the equipment to restore it back to top operation condition.

These all shall be incorporated into a preventive maintenance program which will give automatic time schedule for their routine appliance. This maintenance scheduling shall be the responsibility of the technical services department.

Any one-time major overhaul or group of major repairs specifically planned as scheduled over a limited period of time which should have the effect of service life of the equipment extended two or more years.

7.2   LUBRICATION USE AND SCHEDULE

What is lubrication? Consider two flat plates one resting on the other. Friction will be found to exist between the plates as the resisting forces tangent to the two plate surfaces as they move upon each other. This is called sliding friction. The resistance to motion and degree of wear are very high. When ball or roller bearings are placed between the plates, friction and wear will be reduced and friction here is known as solid friction.

However, friction can be much more reduced if a fluid is placed between the plates. This is fluid friction. If the fluid film between surfaces is sufficiently thick to prevent almost no contact of the surfaces, friction is significantly eliminated and wear is greatly reduced.

Anything introduced between two such surfaces to accomplish a reduction in friction or change the frictional properties is called a lubricant. Lubrication can simply be stated as transforming solid friction to fluid friction with the application of suitable lubricant.

In this case, friction is harmful in bearings, gears and flow of fluids. Therefore, lubricants become to be a part of maintenance for all parts where they are harmful. There are cases when friction can also be useful such as when we use it for brakes, clutches, belts, bolted or riveted joints and self-locking mechanism. This needs some lubrication for control or none at all.

7.2.1   FUNCTIONS OF LUBRICANTS

There are also other purposes of lubricants, and selection and application of lubricants are determined by preventing friction as well as the other functions they are expected to perform. Lubricants vary with respect to their ability to perform the various functions ascribed to them such as lower emissions, improve wear resistance, lower temperatures, improve efficiency or extend engine life.

Important functions are as follows:

Control of friction (main function)	Transmit power (Hydraulic)
Control of wear	Remove contaminants
Control of temperature	Seal
Control of corrosion	

7.2.2 Substance Used as Lubricants

1. Petroleum lubricants
2. Account for 90% of all lubricants
3. Synthetic lubricants
4. Fatty oils

7.2.3 Lubrication System

Practically, all small and a large number of medium speed engines use a single pressure lubrication system combined with splash lubrication in the crankcase. For larger bore engines, it is usually necessary to employ separate cylinder lubrication to insure proper application and distribution of the oil. This is done by the use of a high-pressure mechanical lubricator, usually driven from the camshaft, which delivers all through "quills" to several prints located radically around the cylinder.

All engines have a basic pressure lubrication system which comprises a circulating pump which obtains its supply of oil from either the crankcase oil pan (in a wet sump engine) or remove contaminants, an oil cooler to maintain proper oil temperatures, suitable strainers and pressure-regulating valves to control pressure. Oil is distributed under pressure through piping and drilled passage to the crankshaft bearings, camshaft bearings, piston pin or crosshead, valve mechanism and other moving parts requiring the crankcase lubricant.

In most engines, the crankcase oil is also used to carry away the heat from the under crown of the pistons. However, some large slow speed engines employ water to cool the pistons through the use of telescoping tubes. The cylinders of the large crosshead engines are isolated from the crankcase, while those of the trunk-type engine are not. The blow-by gases and spent cylinder lubricant from the crosshead engine do not discharge to the crankcase, so contamination from this source is not significant.

Thus, the lubricating oil in the circulating system normally has a reasonably long life. However, the circulating oil in a trunk-type engine is exposed to the blow-by gases and cylinder debris and requires greater purification and more frequent oil changes. In a trunk-type diesel engine, acidic products of combustion result in the accumulation of insoluble materials that promote oxidation of the oil. In gas and dual-fuel engine, gas blow-by causes nitration and oxidation degradation of the oil.

7.2.4 LUBRICATING OILS

Fossil crude oil is distilled to come up with different products use such as petroleum, oil and lubricant (POL). After the crude oil is first distilled or fractioned, it shall have removed the lighter volatile hydrocarbons, such as gasoline, kerosene and distillate fuel oil, leaving the heavier oil fraction from which the lubricating oil is produced. By the use of solvent refining, acid treating, filtration, dew axing and additional fractionation, the undesirable crude constitute are further removed, leaving only these hydrocarbons which met the standard requirements of a lubricating oil. From here, different types and grades of lubricants are produced to meet the wide variety of conditions found in engine and machine operating conditions.

7.2.5 ADDITIVES

The petroleum industry has improved lubricating oil quality through improved refining methods. However, the major advance has made through the use of now and highly complex synthetic chemicals. Lubricating oil is required to perform many diverse functions. In addition to its primary functions of minimizing friction and wear, oil must also keep an engine cleaned from burned carbon and free of rust. In these engines equipped with hydraulic lash adjusters, it must also act as hydraulic oil.

Additives are then produced to be supplement lubricating oil to expand the service usage of the lubrication oil. Most of the high-performance oils contain various combinations of these additives, depending upon the service that they are designed to meet.

Physical and Chemical Properties

a) Appearance, odor and color	f) Viscosity index
b) API gravity	g) Carbon residue
c) Flash and fire points	h) Ash content
d) Pour point	i) Total base number
e) Viscosity	

7.2.5.1 Additives according to Usage

Oil additives are proven as vital for the proper lubrication and prolonged use of motor oil in modern internal combustion engines. Without it, the oil would become contaminated, break down, drip out or do not properly protect engine parts at all high engine temperature. Just as important are additives for oils used inside gearboxes, automatic transmissions and bearings.

Some of the most important additives include those used for viscosity and lubricity, contaminant control, the control of chemical breakdown and seal conditioning. Some additives permit lubricants to perform better under severe conditions, such as extreme pressures and temperatures and high levels of contamination.

7.2.5.1.1   Controlling Chemical Breakdown

1. Detergent additives are used to clean and neutralize oil impurities which would normally cause deposits (oil sludge) on vital engine parts.
2. Corrosion or rust inhibiting additives retard the oxidation of metal inside an engine.
3. Antioxidant additives retard the degradation of the stock oil by oxidation.
4. Metal deactivators create a film on metal surfaces to prevent the metal from causing the oil to be oxidized.

7.2.5.1.2   For Viscosity

1. Viscosity modifiers make oil's viscosity higher at elevated temperatures, improving its viscosity index (VI). This combats the tendency of the oil to become thin at high temperature. The advantage of using less viscous oil with a VI improver is that it will have improved low temperature fluidity as well as being viscous enough to lubricate at operating temperature. Most multi-grade oils have viscosity modifiers. Some synthetic oils are engineered to meet multi-grade specifications without them.
2. Pour point depressants improve the oil's ability to flow at lower temperatures.

7.2.5.1.3   For Lubricity

1. Friction modifiers or friction reducers are used for increasing fuel economy by reducing friction between moving parts. Friction modifiers alter the lubricity of the base oil.
2. Extreme pressure agents bond to metal surfaces, keeping them from touching even at high pressure.
3. Anti-wear additives or wear inhibiting additives cause a film to surround metal parts, helping to keep them separated.
4. Wear metals from friction are unintentional oil additives, but most large metal particles and impurities are removed in situ using either magnets or oil filters.

7.2.5.1.4   For Contaminant Control

1. Dispersants keep contaminants (e.g. soot) suspended in the oil to prevent them from coagulating.
2. Anti-foam agents (defoamants) inhibit the production of air bubbles and foam in the oil which can cause a loss of lubrication, pitting and corrosion where entrained air and combustion gases contact metal surfaces.
3. Anti-misting agents prevent the atomization of the oil. Typical anti-misting agents are silicones.
4. Wax crystal modifiers are de-waxing aids that improve the ability of oil filters to separate wax from oil. This type of additive has applications in the refining and transport of oil but not for lubricant formulation (Table 7.1).

TABLE 7.1

Additional Table of Commonly Used Additives

Additive Type	Type of Compounds Commonly Used	Reason for Use	Possible Mechanism
Dispersants	Alky polyamides, Alkyl P255 products, nitrogen containing methacrylate polymers, metal sulfonate, organic boron compounds.	Maintain engine cleanliness by keeping insoluble material in suspension	Primarily a physical process, dispersant is attracted to sludge particle by polar forces. Oil solubility of dispersant keeps sludge suspended.
Detergents	Present detergents are surfactants usually of the metallic sulfonate or phenolate type combined with varying degrees of alkalinity	Provide clean metal surfaces and maintain an alkaline or neutral lubricant.	Surfactant nature of detergent cleans metal surface and acid constituents of the oil are neutralized by the alkaline portion of the detergent.
Viscosity index improvers	Methacrylate polymers, butylenes polymers, polymerized olefins or isoolefins alkylated styrene polymers and various selected copolymers.	To lower the rate viscosity with temperature.	VI improvers are less affected to temperature than oil. They raise the viscosity at 21°F more in proportion than at 10°F due to change in solubility.
Oxidation inhibitors	Zinc dithiophosphates hindered phenols, acromatic amines	Retard oxidative decomposition of the oil which can result in varnish, sludge and corrosion.	Decompose peroxides, inhibit free radical formation and passivate metal surfaces.
Corrosion inhibitors	Zinc dithiophosphates, metal phenolates, basic metal sulfonates.	To prevent attack to corrosive oil contamination bearings and other engine	Neutralization of acidic material and by the formation of a chemical film on metal surface
Metal deactivates	Zinc dithiophosphates organic sulfides, certain organic nitrogen compounds.	Passivity catalytic metal surfaces to inhibit oxidation.	Form inactive protective film on metal surface. Form catalytically inactive complex with metal ions.
Anti-wear extreme pressure (SP) and oiliness, film strength agents	Zinc dithiophosphates organic sulfides, certain organic nitrogen compounds, boron-nitrogen compounds.	To reduce friction, prevent scoring and seizure to reduce wear.	Film formed by chemical reaction on metal-contacting surfaces which have lower shear strength than base metal, thereby reducing friction and preventing welding and seizure of contacting surfaces when oil film is ruptured.

(Continued)

TABLE 7.1 (*Continued*)
Additional Table of Commonly Used Additives

Additive Type	Type of Compounds Commonly Used	Reason for Use	Possible Mechanism
Anti-wear extreme pressure (EP) and oiliness film strength agents	Zinc dithiophosphates, organic phosphates and acid phosphates, organic sulfur and chlorine compounds, boron-nitrogen compounds.	Prevent rusting of ferrous engine parts during storage and from acidic moisture accumulated during cold engine operation. This is a specific type of corrosion.	Preferential adsorption on metal surfaces. This film repels attack of water. Neutralizing corrosive acids.
Rust inhibitors	Metal sulfonates, fatty acids and amines.	To lower pour point of lubricants	Wax crystals in oil coated to prevent adsorption at reduced temperatures
Foam inhibitors	Silicone polymers	To prevent the formation of stable foam.	Produce surface tension which allows air bubbles to separate from the oil

7.2.6 ENGINE OIL TEST FOR ANALYSIS

To verify oil and engine performance after some period, the engine used oil is taken as sample from the engine and sent to a laboratory to analyze performance of primarily the oil and then of the engine. There are two principal reasons for analyzing used engine oils: (1) to determine the used oil's suitability as a lubricant and (2) is to disclose environmental, operational or mechanical problems of the internal parts of the engine in contact with the oil.

For example, the operator of a large, slow speed diesel engine containing many gallons of lubricant is vitally concerned not only with the condition of the oil for its own value but also with early warning symptoms of engine parts premature failure. A fleet operator or a contractor desires optimum drain periods for mobile equipment and needs to know whether his maintenance schedules are appropriate for the working environment. When operational or performance problems cannot be diagnosed on site, an oil analysis in support of "on-the-spot" observations often provides the information necessary for taking corrective action.

Used oil analysis is generally supplied to the customer in a laboratory report that lists test data and offers comments relative to the oil's condition. If adverse factors are indicated, recommendation for their correction is included by the laboratory.

The effectiveness of used engine oil analysis depends on experience and technology when there is agreement on how the properties of used oil are to be measured and explicit definitions of what is being measured; the technology can be dealt with on common ground. Almost anyone who regularly handles oil test data sooner or later develops his own "rule-of-thumb" as to the interpretation.

Such guidelines are the result of actual experience, and they may be applied rigidly or loosely, depending on how well they fit a particular set of circumstances. As in most laboratories, used oil technology provides more than one analysis method and it is therefore understandable that different laboratories do not necessarily use identical approaches.

7.2.7   HANDLING OF PETROLEUM, OIL AND LUBRICANTS (POL)

Satisfactorily lubrication depends upon a combination of several factors. It's not only the necessity of assuring that if it reaches the equipment to be lubricated in the right quantity and at the right time and of the right type, but in addition, there is one other thing which is certainly of equal importance but which is emphasized and stressed all too infrequently, that is the need for proper handling and storage so that when it is applied, it is in the same condition as when it is packaged.

Little does it matter how much is devoted to the selection of the lubricant or fuel, or how efficient and elaborate may be the arrangement for applying it, if at the moment of application, the lubricant is not in suitable condition.

The unfortunate consequences that may result careless and improper storage and handling are much too serious to risk. The least that could happen would be less of the lubricant, if the condition was noticed in time, and it had to be discarded. However, if contamination or any other abnormal condition was not detected before the lubricant was used, the outcome could be very serious and costly, particularly if it caused a lubrication failure on some expensive equipment. In short, mishandling of the products could lead to disastrous results. Handling and storage of POL is a warehouse concern and must be looked into by the maintenance engineers.

7.2.7.1   Guidelines for Safe Unloading of POL

1. Cleanliness is essential regardless of whether storage in indoor or outdoor. For indoor storage, area should be pointed in light colors, preferably white. Maintain regular cleaning schedules.
2. Orderliness is another must. By keeping different brands and types of lubricants separated and in their proper places, there is less chance for confusion and error.
3. Avoid exposure of the lubricants to moisture as content with water may affect same types of lubricating oils and greases.
4. Exercise extreme care in heating products that have been exposed to low temperature. If possible, transfer drums to a warm area and allow product to reach room temperature. Never allow direct heat as sunlight to keep the product integrity. This may destroy sealing compound and cause leakage and might also damage the product.

7.2.7.2   Guidelines for Handling of Lubricants

1. A drum equipped with a pump, hose and control value mounted on a dolly may be devised to eliminate intermediate transfer of lubricants.

2. When many lubricants are used, a simple code system can help avoid confusion, and the code, whether it be alphabetical, numerical or color, should be clearly marked.
3. Keep disposing area clean – don't let spills or drips accumulate as they are safety hazards.
4. Keep partially filled drums or containers properly covered to prevent dirt or dust contamination. Empty containers should never be left in the lubricant storage room.
5. Grease guns should be filled, cleaned and overhauled on a clean bench.
6. Keep fire protection equipment handy and in operable conditions.

7.2.8 LUBE TRUCK

Equipment may be distributed at different project sub-sites or maybe located in one far-out job site. Examples of these are railway projects, when different sites are simultaneously operating, or dam projects when the site is far-out from the central equipment office and yard of the project. In this instance, a lube truck for the regular PM maintenance service of the heavy equipment is the solution.

A lube truck must be well equipped to be able to dispense and fill equipment POL requirement at the project field site with minimum downtime. Some good features are equipped with diaphragm pump and grease pump driven by air pressure from a compressor, hose system are retractable hose reels, centralized board control panel if possible although individual meter may be used but is cumbersome and can carry multiple tanks ranging from 100 to 4000 liters for big units.

The unit must be kept clean for easy and simple maintenance. It should also have a compartment for record keeping and board for writing. Unit must return to the shop at the end of the day for refurbishment.

Some typical compartments are chemical oil, AFT oil, engine oil, Lubricants, emergency diesel fuel, and coolants.

7.3 TIRE CONTROL SYSTEM

Tire cost continues to be one of the largest expenses in the operation of wheeled-type heavy equipment. An efficient tire control and records keeping system is helpful in reducing overall tire cost, where one of the most notorious hidden costs of equipment can originate. With good records, maintenance personnel can control and also identify which brands and types of tires are performing best for a particular project.

Here we describe in detail the tire control and record keeping system. Also included are tips on how to properly maintain tires. Some tips here on inflation, tire matching, mounting, retreading and more will assist in getting the maximum value and life from these tires.

The objectives of the tire control and record system are to be able to know as well the total cost per operating hour or kilometer of tires to aid maintenance management in determining the type or brand of tire giving the best service.

7.3.1 TIRE BRANDING (MARKING) SYSTEM

The key in tire branding or marking system is to identify each tire individually and track its usage by the project. To do this, tires must be properly identified. The use of serial numbers is a logical means of identifying each tire. However, serial numbers are often too long and cumbersome to use easily. Also, tires conceivably have same serial number, particularly tires purchased in large lots. Tire branding is the most efficient method of identifying company tires and the easiness to control tires.

Tires are branded as shown in the example:

Example of a tire brand: **XXX11 – 001**
    Where: XXX is the company or project name
        11 is tire size; 11.0 × 20
        001 is the first 11.0 × 20 tire purchased and increases
consecutively.

The following are the tire brand codes sample for a typical project (Table 7.2):

7.3.2 TIRE CARE AND MAINTENANCE

7.3.2.1 Inflation

Perhaps the most important aspect of tire maintenance is proper inflation. You can have tires marked correctly and mounted correctly, but without proper

---

TABLE 7.2

Brand Number	Tire Size (S)	Brand Number	Tire Size (S)
XXX -6-001	6.50x6; 6.40x14; 6.40/6.50x12; 6.00x12; 6.40/6.50x14; 6.95x14	XXX-14-001	14.00x24
XXX-7-001	7.50x20; 7.50x16; 7.00*x5	XXX-16-001	16.00x24
XXX-8-001	8.25x20	XXX-18-001	18.00x25
XXX-9-001	9.00x20	XXX-20-001	20.5x25
XXX-10-001	10.00x15; 10.00x20	XXX-21-001	21.00x35
XXX-11-001	11.00x20	XXX-23-001	23.5x25
XXX-12-001	12.00x20; 12.00x24	XXX-29-001	29.5x25; 29.5x29
XXX-13-001	13.00x24	XXX-33-001	33.5x33

inflation, the risk of reduced tire life is expected. There was a project instance when a new 33.5 × 33 tire was only used for a week due to wrong inflation and that costed the company much in absence of simple control. The management never found out there was additional cost of such expense due to lack of control.

The following tips help in controlling tire costs through proper inflation:

1. Check the tires with a master gauge regularly. Make sure your gauge is accurate.
2. Inflate recommended pressure for the maximum load to be carried. For long high-speed runs, pressures can be increased by ten pounds as noted on tire specification sheets.
3. Use quality valve caps on all tires. It keeps dirt out of the valve stem and seals off slow leaks. Plastic can melt in high-heat conditions. Use a cap made of heat resistant material.
4. On newly mounted tires, check the inflation again in 24 hours. Tire growth can reduce the original pressure.
5. Never reduce air pressure to get a softer ride. Underinflation causes tires to flex and build up heat. Heat softens rubber and leads to more rapid wear. In cases of extreme heat, build-up tires can catch fire.
6. Don't compensate for overload by increasing tire pressure. Overinflation makes casings more susceptible to impact breaks and uneven tread wear.
7. Never bleed a hot tire. Tire temperature rises while the tire is run limits of air pressure, load and speed, build-up of 10–15 lbs are normal. Anything over that should be corrected immediately.

Note: Over- or underinflation causes tread to wear in areas which are not supposed to be worn out such as the casing. Valuable miles are lost and could cause permanent damage to the casing. This commonly happen without the awareness of the management.

7.3.2.2 Proper Mounting

Correct mounting of tires on their rims is of utmost importance. There are different procedures to be used to mounting tires properly. These are available from most wheel and rim manufactures and distributions.

Here are a few tips to follow to avoid many mounting problems.

- Tires are designed for a specific size and type rim. Use the right tire size for the right rim size. Incorrect load supports cause more rapid tread wear and bead problems. It will end up costing your money.
- Make sure a newly mounted rim or shell is retightened after approximately 150 kilometers. It usually takes that long for any rust or dirt to shake loose.
- Make a visual check of all newly mounted tires.

Check the following for proper mounting:

1. Check flanges for bends or cracks, lock seat is properly sealed and bolts have no cracks.
2. Always use an inflation safety cage when inflating tires. The concussion of sound from an exploding tire can break ear drums, so imagine the force behind anything flying through the air at you.
3. When mounting or dismounting tubeless tires, especially avoid any damage to the bead area. Any chipped or damage bead could cause air to migrate through the casing and eventually cause a separation. It may also have a constant slow leaker on hand.
4. In tube tires always put a new tube in a tire. A tube that has been previously run will have stretched to fit the tire it was in. Also use the right-sized tube. Buying tube on size under may seem like saving money, but it isn't. The stress of trying to fit that tire will often cause a tube to fail, particularly in the tube splice area.
5. Use rim parts that are meant for each other. Using one brand of rim and another brand of flange is only asking for trouble.
6. On tubeless tires, check the valve stem. Often heat and flexing can damage the valve seat and cause a slow leak. Replace stem each time a tire is changed.
7. The last work in mounting is CLEAN. Keep the area where tires are mounted and dismounted free of oil and grease. Make sure all rim surfaces are free of rust and dirt. Rims, tires, and tubes all last lower when they are clean.

7.3.2.3   Tire Selection

There are basically five major points to remember in tire selection. They must be considered together because each affects the other. You may have to compromise one aspect for another to get exactly what you need.

The five major points are as follows:

1. Consider the type of vehicle and the use it is intended for. It must be decided whether to accept the original equipment tires or ask for options. However, option is limited by the vehicles' clearances and recommended rim size. For instance, if the truck is intended for long, over the read operations, the tries should be designed for long life and safety.
2. Consider the loads that will be carried. Get enough tires to support the maximum loads that will be carried. It may not be loaded to maximum each time, but when it is at maximum, it will need a stronger tire. Never exceed the specified load-carrying capacity of a tire however.
3. A third aspect of tire selection is the speed and continuity of operation. For instance, long, higher speed hauls cause a tire to build-up heat. And

heat causes tire failures and rapid tread wear. For long high-speed hauls, look at tires with cooler running characteristics.

4. Consider the type of tread design that best suits the operation. Rib-type designs are best suited, obviously, for front wheel positions. However, they can also be used for drive wheels of tractor/trailers and larger straight truck used primarily over the road. Lug-type design delivers more miles than rib design. They can also be moved back to trailers for additional service. The severe lug design should also be used in slower on/off road conditions and areas where traction is essential, such as in mud.

5. Last, but not the least, buy quality new tires that have a good reputation and for specific each type of area application such as sandy or dessert surface, rocky surface or hard surface.

7.3.2.4   Tire Matching

Mismatched duals cause the larger tire to overload. When a tire is overloaded, the increased flexing of the tire causes rapid heat build-up. The smaller tire will skip or scuff causing uneven, rapid tread wear.

Here are some tips on maintaining matched duals:

1. Try to minimize mismatching by using tires of the same brand and constructions side by side. Different brands could have different degrees of flex built in.

2. Never mate a lug tire and a rib tire on the same dual position. Even if they are the same size, their rolling resistance and traction qualities are much different.

3. Never mate tires of different construction. Radials should be mated with radials and bias ply with bias ply.

4. Maintain proper spacing between duals. This allows air to circulate between tires and keeps tires cooler.

5. When in doubt, check the tire and rim manuals for maximum matching tolerances. Generally, they are ¼ inch for 8.25 Outside Diameter (O.D) cross section and smaller.

Basic tire tools must be readily available for all tire men to check proper matching which includes the following:

1. Steel tape	3. Tire square
2. Tire caliper	4. String gauge

Check tires for proper matching again within 24 hours of mounting. Tire growth in that period may cause a mismatch that wasn't evident earlier.

If you see an irregular wear pattern beginning, check matching tolerances again. Something may have been overlooked in the initial inspection.

7.3.2.5   Driving Tips for All on Highway Equipment

In any tire control system, the assistance and cooperation of the drivers are paramount. If drivers abuse tires, they won't give the performance desired.

Here are a few tips for drivers to be aware of so they can help to get the most out of tires.

1. Probably the best advice to pass on to drivers is not to keep secrets. If a driver notices an irregular wear pattern starting, it should be reported at once and corrected.
2. If a driver feels a tire is not performing like it should, accept his advice on how to correct it.
3. Don't spin the wheels. Every turn of the wheels without traction just scrapes off good rubber and wastes engine power and fuel. And any rocks or hazards that are in the tread grooves are just driven in even deeper.
4. Avoid parking in oil, fuel or grease. Rubber will soak up oil and grease and begin to deteriorate.
5. Don't rub tires against curbs or walls. Even a heavy tire, a truck tire, can be damaged by this practice.
6. If driving within the city, watch for sudden stops and starts. This could lead to flat spots in the tread that will eventually lead to irregular wear.
7. Avoid driving over obstacles whenever possible. You wouldn't deliberately cut a tire, so why take the chance by driving over something you could avoid.
8. Always use recommended pressure for each specific tire size. Keep this information available to all drivers and operators at all times.

7.4   UNDERCARRIAGE CONTROL SYSTEM

Like tires, undercarriage is another important equipment component that should have a control and record system set up. This is a handy reference and record of recommended undercarriage measurement techniques, wear limit tables and the effects of various operational environments on undercarriage life. Included below is the undercarriage control and record keeping system.

Since the undercarriage is the major wear item on track equipment, reliable analysis of wear and prediction of potential undercarriage life is needed to operate at the lowest possible cost and downtime. The best way to provide reliable recommendations is through close follow-up-observe what the tractor is working in and how it's being worked.

The objective of the undercarriage control and record keeping system is to provide maintenance management with undercarriage information regarding what repairs were made, how long they last and what they cost.

7.4.1 UNDERCARRIAGE BRANDING SYSTEM

The key in undercarriage branding system is to treat each undercarriage component individually. To do this, component must be properly identified. Undercarriage branding is the most efficient method of identifying undercarriage component and the easiest to establish.

Brand numbers shall be referred to as property numbers which shall be marked on all undercarriage components on areas where it could be easily noticed and could not be erased by constant usage.

Brand new carrier and track rollers shall be marked by "punching"; subsequently, property numbers shall be welded prior to assembly during rebuilding. The reason for punching property numbers on brand new rollers is to prevent damage of internal seals and o-rings. Other components such as track ass, idlers and sprockets shall be marked by welding though brand new.

Following are examples of progressive property numbering for XXX company undercarriage components:

XXXD6 – 01	(For CAT D6R Tractors)
XXXD7 – 01	(For CAT D7R Tractors)
XXXD8 – 01	(For CAT D8R Tractors)

To effectively illustrate the manner of assigning property numbers, consider the receipt of a CAT D6R (or any model) Tractor. Each undercarriage components according to their quantity per unit shall be marked as follows:

Track Assy.	:	D6-01 & D6-02
Idlers	:	D6-01 & D6-02
Sprockets	:	D6-01 & D6-02
Carriers	:	D6-01 & D6-04
Rollers		D6-01 & D6-12

7.4.2 UNDERCARRIAGE CONTROL AND RECORDS

The undersigned record keeping system involves three basic forms: undercarriage record card, undercarriage inspection report and undercarriage change report. The form is basic and somewhat self-explanatory, but it is best to go through the necessary steps of each form to give a more thorough understanding of how the forms should be completed. Also, these directions may be of help in explaining procedures to maintenance crew directly involved in the undercarriage control.

7.4.2.1 Undercarriage Record Card

It is a form to account for every movement and service performed by or on the undercarriage component from the start to the end of its usable life.

Name of Company and Address

Undercarriage Record Card

Equipment No.		Date Acquired:
Make & Model:		
Description:		
Inspection Remark:	OK - Good Cond.	RH – Rehab
	R - Repair	RP – Replace Part

Components	Pcs. Left(L)	Property No.	Inspection Remark	Action Taken	Components	Pcs. Right(R)	Property No.	Inspection Remark	Action Taken
Track Assy.	L				Track Assy.	R			
Idler Assy.	FL				Idler Assy.	FR			
Carrier Assy.	RL				Carrier Assy.	RR			
Carrier Assy.	L1				Carrier Assy.	R1			
Roller 1	L2				Roller 1	R2			
Roller 2	L3				Roller 2	R3			
Roller 3	L4				Roller 3	R4			
Roller 4	L5				Roller 4	R5			
Roller 5	L6				Roller 5	R6			
Roller 6	L7				Roller 6	R7			
Roller 7	L8				Roller 7	R8			
Roller 8	L				Roller 8	R			
Sprocket Assy	L				Sprocket Assy	R			
Pins & Bushing (ext)	L				Pins & Bushing (ext)	R			
Pins & Bushing (int)	L				Pins & Bushing (int)	R			
Track & Link					Track & Link	R			
Track Shoe					Track Shoe				

7.4.2.2   Undercarriage Inspection and Change Report

The above report card can be used to report for the present physical status, condition and replacement of undercarriage components and to provide maintenance management with data and information vital to undercarriage planning and forecasting. More columns can be added to details of dates of removal and replacement of assembly per equipment.

7.4.3   UNDERCARRIAGE MAINTENANCE PROCEDURE

7.4.3.1   For Maintenance Men

1. Check the working clearance between roller flange and pin boss. As link rails and roller treads wear, clearances at the inner side of the flange disappear first. Visual inspection will not normally show contract between the flange and the pin boss, here. It may look as though there is still sufficient clearance when the roller flange is actually wearing against the pin boss. But as roller flange-pin boss contact begins, a very definite wear pattern will appear on the top of the pin boss. Contact between the two will speed roller flange wear.

2. Adjust track so that it sags about 1¼ inch to 13/4 inch between the front carrier and the idler. If the track is too tight, damaging stresses are placed upon it. Track wears out more quickly and tractor drawbar horsepower is reduced. On the other hand, when track is too loose, it whips at high speed, causing links to crash against carrier rollers.

3. Keep track hardware tight. Loose hardware will cause bolt holes on shoes and links to wallow out. It can also result in hardware failure and loss of track shoes. Use square nuts on links with self-locking seats. Place chamfered (rounded) side of the nut against the bolt seat.

4. Examine idler flanges for excessive wear. Wear on just one side may mean the idler is off center. Correct with a shim adjustment. Excessive wear on both sides means the wear strips on the shims are worn out. On some tractors, adding shims may correct the problem while on other new wear strips may be needed. Another possible cause of this type of wear is because roller frame could be out of alignment.

5. Tractor with idler petitioners should be operated with idler in high position whenever possible. This makes turning easier and puts less strain on track. Low position may be necessary, however, when full stability and ground contact are needed, particularly with machines equipped with dozer blades.

6. Hill-side work can cause excessive wear on roller flanges and link sides. Track guiding guards are the best defense against this type of wear. If machine engaged in side hill work for a prolonged period of time, condition of the guides and wear strips should be checked regularly. Keep in mind that guiding guards are a good investment in most applications.

7. Watch for signs of internal pin and bushing wear.

8. Dozing and push loading cause faster wear on front track rollers. If flanges and treads on front rollers are wearing noticeably faster than on other rollers, determine whether it would be economically sound to switch front and rear rollers to balance the wear.

7.4.3.2 For Equipment Superintendents

1. Be familiar with various undercarriage options available from Caterpillar dealers. These may help reduce costs or improvements performance of equipment. For example, optional 6.91-inch pitch track for D6 Tractors will provide longer wear life in severe applications. Optional 9.00-inch pitch track for non-current D8 Tractors is heavier and lasts longer in most applications than the standard 8.00-inch pitch track.

2. Keep good records. The only way to know how undercarriage is performing is to know what repairs were made, how long they last and what they cost.

3. Be alert to the symptoms of internal pin and bushing wearing loose "snaky" track. Extended track pitch, resulting from internal pin and bushing wear, accelerates wear on rollers, idlers, links and guiding guards. Remember that all components of the undercarriage system are interconnected. Internal pin and bushing wear will set off "chain reaction" of wear on other parts of the system.

4. Guiding guards are a good investment in any operation. These are vital in operations where excessive side wear on links, rollers and sprockets occurs. For example, if a machine does a lot of side hill work or is constantly turning, guiding guards will normally retard side wear on links, sprockets and rollers flanges. To further reduce maintenance costs, inexpensive bolt-on wear strips are available for guiding guards on many tractor models.

5. If a machine is experiencing excessive wear on one or the other, consider rotating the sprockets. If the machine is equipped with bolt-on sprocket segments, they may be removed and installed on the opposite side of the machine to retard excessive wear on one side of the teeth.

7.4.3.3 For Operators

1. When a tractor is operated in rough, rocky materials, keep in mind that the track takes a severe pounding. Sometimes it is necessary to "push" rig to maintain a production schedule, but sustained and needless abuse shortens track's life. A little consideration for the machine will cut maintenance costs and keep the unit on the job.

2. Make daily visual inspection of the equipment. Check for loose bolts, leaking seals and abnormal wear. Report all items that need attention to the maintenance man so that needed adjustment can be made before damage occurs.

3. Avoid unnecessary speed. It quickly wears out undercarriage parts over rough ground, high-speed damages and mis-aligns front track rollers and idlers. Speed shortens undercarriage service's life.

4. Never park a machine on the side of a hill. This puts a strain on rollers seals and oil may leak out. Next morning, it will take time before the rubber rings return to the correct shape. In the meantime, even more oil is lost. Rollers only hold about one pint of oil. Without it, the internal parts of a roller will fail prematurely.

5. When operating a dozer or scraper, don't take such a deep cut that tracks begin to spin (dozer operators using hydraulically controlled blades sometimes overuse down-pressure, lifting the front end of the machine into the air). When the tracks begin to spin, no work gets done but tracks' shoes wear out at a faster rate.

6. Back up slowly. High-speed reverse is the most deadly enemy of pins and bushings. Reason: gins and hushing are put under greater strain when the tractor is backing up. Also, when speed increases, wear increases faster.

7. Working in a pattern that requires constant running to the right or left will wear out one track faster than the other. If job conditions permit, change of operating direction will help balance wear.

7.4.4 Undercarriage Maintenance Checklist

The following is a brief summary of maintenance and operating recommendations that will contribute much to getting longer track life. Remember, however, that these are only the basic steps. Maximum savings can result only from following the recommendations of the manufacturer operator's manuals

1. Slowdown. Don't go fast unless productivity is worth the increased wear. This is especially true for high-speed reverse.

2. Maintain proper track alignment. A properly aligned track roller frame adds life to rollers, idlers, sprockets and links.

3. Adjust track for proper sag. Measure the sag while the machine is in actual working condition to ensure accurate result.

4. Check rollers of oil leaks. The oil must be dyed red and so can be seen easily.

5. Clean mud and debris from undercarriage so rollers can turn properly. Also, it'll be easier to check for leaks.

6. Be sure that idlers are in the proper position. Generally low position is for dozing and high position is for drawbar applications.

7. Use track guiding guards to reduce roller flange and link side wear, and extend undercarriage's life.

8. Always use the narrowest shoe possible which still gives you adequate floatation. Don't spin the track. A spinning track severely increases grouser wear.

9. Tighten track hardware properly. Loose hardware can cause wallowed-out holes, lost shoes and links which cannot be rebuilt.

10. Turn pins and bushings. If done properly, service life can be increased up to 50%.

11. If you have belt-on sprocket segments and sprocket wear is mostly on one side, reverse for more wear life.

8 Maintenance Job Order System

Maintenance job order system is a means whereby every type of job done, whether minor or major, is uniformly prepared to show the work to be done, description of work is complete and clear, and details of the equipment data are readily available.

The system is preceded by a job number system for control and information on the equipment repair record. This will facilitate decision and actions needed in time when an equipment status and history is paramount. Example of a major and minor repair list is given below.

Typical work item for major repair of a crawler tractor (CT):

1. General/inframe engine overhaul.
2. Transmission overhaul.
3. Repair drive line (power train).
4. Repair final drive.
5. Repair brake system (hydraulic component).
6. Repair/overhaul hydraulic system.
7. Repair hydraulic component of steering system.
8. Repair undercarriage components.
9. Repair or replace hydraulic cylinders.
10. Repair oil cooler and pumps.
11. Injection pump calibration.
12. Fuel nozzle/injector calibration.
13. Repair hydraulic pumps and valves.
14. Repair/overhaul torque converter.
15. Repair radiator assembly.
16. Repair cooling system, water pump and thermostat.

Typical work item for minor repair of a CT:

1. Repair/rebuild moldboard assembly.
2. Re-harness electric connections.
3. Repair/replace air, water, oil, fuel lines and hose fittings and eliminate leakages.
4. Repair/rebuild engine mounting and accessories.
5. Welding job on cracked and dilapidated floorings.
6. Repair/replace defective control and gauges.

DOI: 10.1201/9781003360667-11

8.1 EQUIPMENT JOB ORDER PROCEDURE

The job order form is the main output of the Job Order system. It is a maintenance form to contain all data of equipment repair, materials and labor contribution to repair equipment. In general, it covers major repairs, general overhaul, conversion, modification, fabrication, construction and those repairs described in 5th echelon maintenance and the unscheduled major repair (non-PM maintenance) of equipment.

The job order shall show that the work is needed; repairs to be done are technically (i.e. engine, undercarriage, transmission and rear or front axle) described and to be part of the equipment record of ownership.

Job Order control is a means whereby every type of maintenance job done, whether minor or major in scope, is definitely covered by a standard written form which shows what work is needed, adequately described, properly approved and issued by proper shop authority and provides a continuing record of maintenance of equipment.

As in PM, inspection of the equipment is first executed by the equipment inspector to determine the work needed by the equipment. This inspection report (see Part IV) shall serve as the parameters to which the shop technicians will start the repair. Any more eventual repairs shall be reported to the shop supervisor or superintendent to complete the job order description of repair operation on the equipment.

Company Name & Address	Form No._____			
Equipment Management Department				
Job Order Form				
Job Order	Date:_____			
Type of Repair: **Major:** [ ] **Minor:** [ ]				
Equipment No.	**Job Order No.**			
Model/ Serial No.	Priority No.			
Engine Make	Estimated Man-hours:			
Model Serial No.	Estimated Time of Completion:			
Reference (EDSL No.)	Last project from:			
Inspected By:	Last assigned Operator/Driver:			
Main Complain:				
Sequel No. (Shop Works)	Operation	Shop Section	No. Of Men Required	Estimated Man-hours
Shop Planner/Clerk:	Superintendent Approval:			
Shop Supervisor:	Shop Manager Approval:			

Job Order Completion and Cost Summary Report (Shop Report)
Supplementary Form Attached to Completed Job Order Form

Job Order No:						Date Of Completion:		
Labor Cost			Material Cost			POL Cost		
Daily Job Ticket	Total Man-hour	Total Cost	Warehouse Slip	Material	Total Cost	POL Slip	Material	Total Cost
Total:			Total:			Total:		

Shop Grand Total:		
Outside Source Work Total:	Supplier1:	
Overall Total Cost:	Supplier2:	
Bookkeeper/Audit Clerk:	Date:	

The job Order numbering system usage will readily be identifiable numbering for the type of work if major or minor, the year and month of work, and the succession or sequence of jobs presiding each Job Order for control and electronic database.

Sample Job Order No: Ma or Mr - 122013 - 0001 - Job Site Code – XXXX

<Major(Ma)–Minor(Mr)>-<Month & year>-<Number of job order>-<site code>-<EDP assigned number>

8.2 EQUIPMENT DAILY JOB TICKET

The daily job ticket is a shop form to document the daily maintenance and repair manpower availability and utilization. The record will also show the man-hours expended by each operation phase of any Job Order.

First as all data for maintenance forms, it shall contain the basic equipment data as equipment no., model/make, serial number (also particularly for engine overhauling section work), last operator/driver to have used, last site/project assignment.

Repair data must have the detailed description of work to be done that date, number of technicians and their code number, number of hours of work of each technician, pay code of technicians, night shift code, to total labor cost. Pay code and total are to be supplemented by bookkeeper. Shop base data are to be certified by the shop supervisor and superintendent.

The daily job ticket must be made out daily and submitted by Foremen to the shop planner at the end of the day. Another daily job ticket will be used the next repair day until the repair shall have been completed. All daily job ticket shall be cumulated to record the total working time on the equipment. This shall represent the total labor cost of repair for the equipment.

Daily job ticket can also be used for the evaluation of the shop performance, manpower efficiency and sufficiency, machinery requirements, number of repairs daily and shop overall diligence to handle the repairs of equipment per project or at the central shop.

Company Name & Address									
Equipment Management Department									
Daily Job Ticket						Date: _____			
Shop Section:				Equipment No.					
Job Order No.				Make/Model:		Serial No.			
				Engine Make/Model:		Serial No.			
Technician Name	Employee No.	Hours Worked			Work Performed	Technician Initial	Pay Code	Night Shift Code	Total Labor Cost
		From	To	Total					
Section Foreman:				Shop Supervisor/Superintendent:					

8.3 EQUIPMENT LIGHT JOB ORDER

The Light Job Order is a maintenance form to document a shop order covering the following:

1. Minor repairs described in 1st–3rd echelon maintenance. Approval must come initially from the PM engineer.
2. Preventive maintenance services and repairs to recover or salvage assemblies, subassemblies and component parts. With description of the maintenance activity required to restore the assembly to an operational status. Approval must be from the superintendent.
3. Dis-assembled assembly or sub-assembly which may be from equipment under another major job order at the same shop location or from equipment located at the job site. In this case, the major Job Order must be made reference for proper equipment record and bookkeeping.

The Light job order form shall adequately describe the work needed. In all instance, it must mention specifically the equipment number where the assembly was taken off. It shall also mention the main job order to which the light Job Order must be added to

Copies of the Light job order must be distributed as follows:

1. Project/Shop of Origin – if after repairs, the assembly will be shipped back to requesting project or to a job order.
2. Storage – if after repairs, the assembly will obligated to warehouse stock.
3. Junk – if in the process of repair, the assembly was found to be uneconomical to restore to serviceable condition.

Company Name & Address		
Equipment Management Department		
Light Job Order		Date: _____

Equipment No.	For Assembly: (Description)	
Make/Model	Make/Model:	
For PM: (Indicate PM Number)	Serial No.	

Job/ Repair Description:	Job/Repair Performed:	Number of Technician:
Requested By:	Shop Accepted By:	Prepared By:
Date Started:	Date Finished:	Approved By:

Discharge Recommendation:		
Return to Original Equipment:	Storage:	For Disposal as Junk:

8.4   EQUIPMENT REPAIR RECORD CARD

The Equipment Repair Record Card is a maintenance form intended particularly for vehicular equipment, construction equipment and plants machineries but may be used for other facilities as well, if applicable.

The form is designed to enable maintenance scheduler to be updated with nature and frequency of repairs performed on the equipment at various times in its operating life. It also provides the cost of repairs performed on the equipment for decision making, i.e., rehabilitations, trade-ins or replacement decisions.

The form shall be prepared, maintained and updated by the shop engineer or clerk under the direction and supervision of the shop superintendent.

Name & Address of the Company Equipment Management Department Equipment Repair Record　　　Date File Started: _____								
Equipment No.								
Make/ Model:					Serial No.			
JO/LJO Nos.	Reference	Date Started	Date Finished	Total Man-hours	Labor Cost	Parts Cost	Total Cost	Nature of Repair
Prepared By:								

8.5　EQUIPMENT JOB ORDER PRIORITY TAG

Job order Priority Tag is issued only to equipment undergoing repairs at central shop that are urgently needed by the project. Equipment shall bear "priority tags" to alert all concern of the urgency and seriousness of project demand. Status of maintenance activities of equipment under a priority tag system must be checked to ensure it meets the deadline given to finish repair of equipment. The priority tags shall be color coded and displayed in front right window shield of the equipment. If this is also under repair, it shall be at any convenient front location where it could be noticed such as the rear view mirror of the driver/operator.

The Job Order priority shall be conveyed or determined by the equipment manager and information relayed to the whole shop operation, most especially to the sections involved with the maintenance work of the equipment.

8.5.1　DEFINITION OF PRIORITIES

- Priority 1-job order to take precedence over all other maintenance work and full support be given in the shop floor, labor and parts allotment.
- Priority 2-job order of those maintenance tasks which it is desired to complete as soon as possible and with one week commitment to be finished.

TABLE 8.1

Maintenance Priority Table

Maintenance Job Priority Number and Color Code

Priority 1	Red	Must be finished immediately
Priority 2	Orange	Must be finished within a week
Priority 3	Yellow	Must be finished within a month
Priority 4	Blue	As Scheduled
Priority 5	Green	Waiting for parts
Priority 6	White	As Scheduled – PM Service – do not delay

- Priority 3-job order of those maintenance tasks which are desirable, but which may be completed within one month. In effect, these include maintenance work which must be also without delays.
- Priority 4-job order of equipment with scheduled maintenance work.
- Priority 5-job order of equipment on PM services.
- Priority 6-job order of equipment on repair and is waiting for parts delivery.

The tags are changed from a lower priority to the next as time progresses. Tags are to be designed from aluminum plates 6 inch by 12 inch (gage 30 plate). Each painted with the assigned color code above. Edges must be smoothened and corners must be round finished. The tags must be kept at the office of the shop superintendent.

8.6   EQUIPMENT PARTS TRANSFER REQUEST

The Parts Transfer Request is a maintenance form to authorize and document transfer of part(s) from breakdown equipment to another breakdown equipment. It must be preapproved by the Shop Manager. The transfers shall only be authorized on extreme emergency cases; i.e., the service of breakdown equipment is urgently needed to meet contractual commitment.

Important considerations to justify this procedure are time element, location and cost. Requisitions shall be made to replace the parts transferred to ensure non lost of an asset. If the donor equipment has been junked, no justification is needed, but the valuation of the donor equipment must be adjusted accordingly.

Company Name & Address Equipment Management Department Parts Transfer Authorization					Dated:_____	
On Repair Equipment				Donor Equipment		
Make/Model		Serial No.		Make/Model		Serial No:
Job Order No.				Job Order No.		
Item No.	Quantity	Unit	Part Number	Assembly Group		Description
Prepared By:			Recommended By:		Approved By:	

9 Company-Provided Maintenance Tools and Equipment

The company must provide tools and equipment to perform maintenance and material supplies to the equipment. These tools must be physically traceable from the accounting records to the custody of authorized tool keeper who shall be responsible for their proper use and safe custody. A section of the shop must be provided for tool keeping or tool room.

Controls are aimed to preclude the possibility of losses due to lack of tool to repair equipment. This is to ensure repairs and prevent downtime due to shortage or proper specific tools for the repairs.

The listing attached is provided to guide all operating groups of their project requirements.

9.1 SHOP TOOLS FOR MECHANICS AND OTHER EQUIPMENT

The company-provided shop tools are those provided by the company and are kept at the tool rooms in the maintenance shop (Table 9.1). These are of general utility and issued on a "tool tag system" to whoever are in need of them at the time. These tools are different from those included in the mechanic's standard tool box (List 1).

- List 1 – List of mechanic's standard personal tools.
- List 2 – List of company-provided tools.
- List 3 – List of machine shop equipment.
- List 4 – List of maintenance support equipment.

9.2 TOOL PHYSICAL CONTROL

The tool system is a method used for releasing tools to borrowers in lieu of the signed method. Each mechanic is issued five tool tags of the same control number. The mechanic's name and the tool tags number shall be logged and kept by the tool keeper. The tool tags shall be traded with the tools borrowed. One (1) tool tag is equivalent to one (1) tool borrowed. The tool keeper shall place the tag on the tool room tag board, where the tool was taken. All tools are borrowed during the shift. All tools not returned at 10:00 a.m. of the following day shall be subject for inquiry.

DOI: 10.1201/9781003360667-12

TABLE 9.1
List of Company-provided Shop Tools

List 1 Mechanics Standard Tools

		1. Senior Mechanics Tools
1	Set	Socket wrench 3/8" to 1-1/8" – ½" square drive complete with power handle, ratchet handle and 2 each extension 4"–8" long
1	Set	Socket wrench 8 mm–30 mm – 1/2" square drive
1	Set	Socket wrench ¼"–¾" – 3/8" square drive complete with power handle, speed handle ratchet handle and 2 each extension 4" and 8" long
1	Set	Combination wrench 3/8" to 1-1/8" heavy duty
1	Set	Combination wrench 8mm–30 mm heavy duty
1	Set	Allen wrench 3 mm–19 mm
1	Set	Adjustable wrench 12"
1	Each	Ball peen hammer 2 lbs
1	Each	Ball peen hammer ¾ lbs
1	Each	Feeler gauge .0015"–.035"
1	Each	Feeler gauge 0.038 mm–1.190mm
1	Each	Fliers long nose 6"
1	Each	Flier slip joint 10"
1	Each	Vise grip 10"
1	Each	Diagonal pliers 6"
1	Each	Screwdriver 4" heavy duty
1	Each	Screwdriver 8" heavy duty
1	Each	Philips screwdriver 6" blade 2
1	Each	Hacksaw 12"
1	Each	Tape rule steel 3 m long
1	Each	Crowfoot bar 16" long or pinch bar
1	Each	Cold chisel ¾" edge
1	Each	Punch 3/8" point
1	Each	Center punch
		2. Junior Mechanics Tools
1	Set	Socket wrench 3/8" to 1-1/8" – ½" square drive complete with power handle, speed handle, ratchet handle and 2 each extension 4"and 8" long
1	Set	Socket wrench 8 mm–30 mm – 1/2" square drive
1	Set	Combination wrench 3/8" to 1-1/8"
1	Set	Combination wrench 8 mm–0 mm
1	Set	Allen wrench 1/8"–5/8"
1	Set	Allen wrench 3 mm–19 mm
1	Each	Adjustable wrench 12"
1	Each	Ball peen hammer 2 lbs
1	Each	Ball peen hammer 3/4 lbs
1	Each	Feeler gauge .0015–.035
1	Each	Feeler gauge 0.038mm–1.190 mm
1	Each	Flier, long nose 6"

(Continued)

TABLE 9.1 (*Continued*)
List of Company-provided Shop Tools

1	Each	Flier, slip joint 10"
1	Each	Vise grip 10"
1	Each	Diagonal cutter 6"
1	Each	Screwdriver 6" heavy duty
1	Each	Philips screwdriver 6" blade # 2
1	Each	Hacksaw 12"
1	Each	Tape rule, steel 3 m
1	Each	Crowfoot bar 16" long or pinch bar
1	Each	Cold chisel ¾" edge
1	Each	Punch 3/8" point
1	Each	Center punch

3. Shop Hand Tools

1	Set	Combination wrench 3/8"–1"
1	Set	Combination wrench 8 mm–26 mm
1	Set	Socket wrench 3/8"–1" – ½" square drive complete with power handle, ratchet handle, and 2 each extension 4" and 8" long
1	Set	Socket Wrench 8mm–26 mm – ½ square drive
1	Each	Adjustable wrench 12"
1	Each	Ball peen hammer 2 lbs
1	Each	Flier, slip joint 10"
1	Each	Screwdriver 6"
1	Each	Cold chisel ¾" edge
1	Each	Center punch

List 2 Company-provided Tools

1. Electrical Section

1	Unit	Multi-Meter Tester Wireless complete with standard attachment
(1)	Unit	Soldering Gun, piston grip type, 230 V, 60 cycles, single phase, 300 W
(1)	Unit	High Speed Soldering Iron, 230 V, 300 W.
(1)	Unit	Battery charger, 230 inputs V, 60 cycles, single phase, 6/12/24 V batteries, 100/90/45 A charging rate
(1)	Unit	Thermo battery hydrometer
(1)	Unit	Battery electrolyte filler 2-1/4 gts. capacity
(1)	Unit	Battery cell tester
(1)	Unit	Growler, 220 V, 60 cycles complete with growler adapter for testing small armature
(1)	Unit	Plier, Lineman 7" with Insulation
(1)	Unit	Plier, Lineman 9-1/2" with Insulation
(1)	Unit	Plier, needle nose with side cutter 2-1/4" jaw length, serrated tips
(1)	Unit	Screw driver, Std. Tip ¼" – 15/32"
(1)	Unit	Screw driver, Philips. 2-3/4" – 10-13/16"

(*Continued*)

TABLE 9.1 (*Continued*)
List of Company-provided Shop Tools

2. Preventive Maintenance Section

(2)	Units	Volume pump, 30 l capacity
(2)	Units	Rotary transfer pump, hand operated, to fit 210 l std. drum, "Graco" Model 226-290 complete with bung adapter, 5 ft. – ¾" hose and spout or equivalent
(4)	Each	Grease gun, lever type, complete with hose and coupler, 1 capacity
(2)	Each	Oil dispensing container with flexible spout, 1 gallon capacity.
(2)	Each	Can oiler, piston grip, 16 oz. capacity with flexible outlet.
(1)	Each	Filter strap wrench "CAT" PN-2p8250
(6)	Pcs.	Oil faucet ¾" to fit 400 lbs drum
(2)	Each	Wrench, adjustable 6"
(1)	Set	Wrench, combination 3/8" to 1-1/8", light duty
(1)	Set	Wrench, combination 8 mm–30 mm, light duty
(1)	Unit	Steam cleaner, 150 GPM capacity, electromagic 1600
(1)	Unit	Fil-T-Vac cleaner, 489 CFM, 2 H.P. motor 220 V, 60 cycles
(1)	Unit	Typewriter, manual
(1)	Unit	Calculator, electronic, pocket size

3. Welding Section

(2)	Units	Welding Machine, 250 A, complete with standard attachments and accessories
(1)	Unit	Welding and cutting outfit heavy duty complete with std. attachments and accessories
(1)	Unit	Anvil block, 60 lbs
(2)	Each	Mask, welding
(1)	Each	Goggle
(1)	Each	Torch lighter

4. Painting Section

(1)	Unit	Gardner-Denver air compressor, 22 cfm, electric motor driven, C 5 H.P., single phase, 230 V, 60 cycles, complete with standard attachments and accessories equivalent
(1)	Each	Spray gun, 18 complete with standard attachments
(1)	Roll	Air Hose, 3/8" * 30" L complete with fittings
(3)	Each	Paint brushes, 2", 3", and 4"
(3)	Each	Lettering brush, ½", 5/4"

5. Tire Section

(1)	Unit	Vulcanizer, electric operated, 230 V, 60 cycles, single phase, 350 W, Model T-990 or equivalent
(1)	Unit	Crocodile jack, 10 tons capacity
(1)	Unit	Tire inflator PCL No. 5014-18 complete with 30 feet air hose, with twin chuck
(1)	Unit	Tire gauge, 0-100 psi
(1)	Unit	Tire tread depth gauge, 0-1" pocket type "DILL"
(1)	Unit	Tire changing stands – handles standard rim diameter 13" through 17" and wheels 3-1/2" through 7-1/2" wide

(*Continued*)

TABLE 9.1 (*Continued*)
List of Company-provided Shop Tools

(1)	Unit	Tire inflation hose, 3/8 I.D. 70 feet long, complete with gauge air chuck
(1)	Each	Tire bead remover
(1)	Each	Wrench, rim
(1)	Each	Rim tools 19" long
(1)	Each	Curved tire removing tool 1-3/8" w * 18" L
(1)	Each	Hammer, sledge, 5 lbs
(1)	Each	Hammer, sledge, 10 lbs
(1)	Each	Hydraulic tire remover "Good year" TO-1600
(1)	Each	Giant tire tools, "Goodyear"

6. Mechanic and Shop Tools

(1)	Set	Combination wrench, hexagonal, sizes ½" to 1-1/2", H.D.
(1)	Set	Combination wrench, hexagonal, sizes 10mm–30 mm, H.D.
(1)	Each	Adjustable wrench, 12"
(1)	Each	Adjustable wrench, 15"
(1)	Unit	Bench grinder, 1. H.P., 230 V, 50/60 cycles, 900/3600 rpm, complete with 10" diameter, grinding stones, cord and magnetic switches
(1)	Unit	Portable electric drill, ¾" chuck capacity, 230 V, 60 cycles, single phase
(1)	Unit	Portable straight grinder 6" capacity, slow speed, 220 V, 60 cycles
(1)	Unit	Bench vise, swivel type, 10" max. opening
(1)	Unit	Hydraulic jack, 5 tons capacity
(1)	Unit	Hydraulic jack, 10 tons capacity
(1)	Unit	Hydraulic jack, 50 tons capacity
(2)	Each	Sledge hammer, 5 and 10 lbs capacity complete with handle
(1)	Unit	Lever block, 1 ton capacity, ratchet puller action
(1)	Unit	Chain block, 2-1/2 tons capacity, rigger type, heavy duty, all steel, chain hoist
(2)	Each	Pipe, wrench, straight type, 12" and 14" "rigid"
(1)	Unit	Drill bit, 1/8" to 1-1/4" diameter straight shank
(1)	Unit	Socket wrench, ¾" square drive, ½" to 1-1/4" complete with power handle and 80 mm extension bar
(1)	Unit	Socket wrench, ½" square drive, 8 mm–22 mm complete with power handle and 80 mm extension bar
(1)	Set	Torque meter, 0–350 ft – lbs ¾" square drive with movable pointer and dial
(1)	Each	Torque meter 0–150 ft – lbs ½" square drive with movable pointer and dial
(1)	Each	Lower handle, ½" and ¾" square drive, 16" length
(2)	Each	Ratchet handle, ½" and ¾" square drive, 12" length
(2)	Each	Hand tap, "NPT" 1/8"–½"
(1)	Set	Allen wrench, 1/8"–¾"
(1)	Set	Hacksaw, 12" with blades
(1)	Each	Smooth file, flat, ¼" * 31/32" * 10" L
(2)	Each	Bastard file, ¼" * 31/32" * 10" L
(1)	Each	Round file, ½" Ø * 12" L
(1)	Each	Half-round file, 7/8" * 9/32" * 10"L

(*Continued*)

TABLE 9.1 (*Continued*)
List of Company-provided Shop Tools

(1)	Each	Medium duty valve lifter for 7/8" to 1-3/8" spring diameter "Snap-on" JT-18"
(1)	Set	Metric hexagonal tap and die tools sets, 3 mm–20 mm range complete with standard attachments and accessories
(1)	Set	Tap 7 die combination sets, NC and NF threads, ¼"–¾" complete with standard attachments
(1)	Set	Tube cutting and flaring tools, handles 3/16"–½ outside tube diameter complete with cutter, flaring bar and rod handle, yoke and 5-adapters 3/16", 3/8", and ½"
(2)	Each	Standard thread restorer files, 8 tools in one for 11, 12, 13, 14, 15, 16, 20, and 24 and for 9, 10, 12, 16, 20, 27, 28, and 32 thread/ inch
(1)	Each	Extension bar, ½" square drive, 6" long
(1)	Each	Extension bar, ¾" square drive, 6" long
(1)	Set	EZY out square screw, extractor for sizes ¼", 5/16", 3/7", 7/16", ½", 9/16",5/8", and ¾"
(1)	Each	Standard medium, 3-jaw screw type puller, 9-1/2" max. jaw spread, 5-1/2" jaw reach "Snap-on" # CG-273A
(1)	Unit	A-Frame, 5 tons capacity (to be fabricated)
(1)	Each	Vise grips "Snap-on" Model VP-10WR, 1-5/7 jaw
(2)	Each	Flashlights, 3 battery
(1)	Each	Snip, 10" long, 2-1/4" blades "Snap-on"
(1)	Assy.	Pipe cutter, heavy duty, "Rigid" 1/8"–2"
(1)	Unit	Pipe vise, bench yoke, "Rigid" 4" capacity
(1)	Unit	Pipe threaded "Rigid" 4" capacity
(1)	Set	Socket wrench, extra heavy duty standard set, 1" sq. drive 1-1/4" to 3-1/8" with ratchet heat, sliding bar and 2 each 4" and 6" extension bar
(1)	Set	Inside micrometer 0-1", 0-2", 1-3", and 2-6"
(4)	Pcs.	Hand held valve lapping tool (small vacuum rubber cap) with handle
(1)	Each	Tachometer, hand type speed indicator, 0–10,000 rpm complete with extension bar pointed and concave wheel tip
(1)	Each	Nitrogen charging Kit, PN-755437

List 3 Machine Shop Equipment

(1)	Unit	Lathe machine
		Swing overhead – 19"
		Swing overhead – 24"
		Length of bed – 7"
		H.P. of motor – 7.5, 230 V, 60 cycles, 3 phase.
(1)	Unit	Drill press – variable speed, 150–2000 rpm, 1-1/2" capacity. Powered by 1-1/2 H.P. motor, 230 V, 60 cycles, 3 phase, complete with rapid action drill press vise 6" * 2-1/2" jaw., 6/3-4" max. opening
(1)	Unit	Hose cut-off and stripping machine "Aeroquip" Model C, ¼"–3" cutting capacity, ¼"–2" stripping capacity, 230 V, 60 cycles, single-phase electric motor
(1)	Unit	Hose assy. machine "Aeroquip" Model FI1013 complete with standard attachments

(*Continued*)

TABLE 9.1 (*Continued*)
List of Company-provided Shop Tools

List 4 Shop Support Equipment		
(1)	Unit	Tire truck
(1)	Unit	Lubrication truck complete lubrications outfit (see PM Lube Truck)
(1)	Unit	Mechanic service truck
(1)	Unit	Service pick-up or Toyota Land Cruiser for Equipment Superintendent's service
(1)	Unit	Fuel truck, 8,000–10,000 l capacity complete with power discharge hose and meter
(1)	Unit	Hydraulic wrecker truck 7-1/2 ton capacity
		End of List

Lost and damaged tools should be made known immediately to the tool keeper who shall conduct the inquiry, if necessary, of the actual causes. A memo, Lost/ Damaged Tools Report, should be initiated to Accounting Department charging personnel account of the employee for the following causes:

- Lost due to negligence – 100% costs of exact replacement
- Misuse – 50% of cost
- Damage – 50% of cost

A copy of the memo should be furnished the Employee 201 file of the employee in the Personnel Department for processing at the due time.

9.3 TOOLS PURCHASING AND ACCOUNTING

Adequacy of tools available for use in operation is the responsibility of the Shop Head who must justify the needs to the Equipment Department Head. Final approval, however, of the type and quantity of shop tools shall be provided by the Equipment Department Head. In case a request is made for a replacement of lost or damaged tools, the requisition must attach a Report of Lost/Damaged Tools to be requisitioned.

9.3.1 ACCOUNTING FOR TOOLS

The tool keeper is responsible for conducting annual physical inventory of tools charged to their responsibility. This includes tools specially and temporarily issued to project job sites within the project, if any under their supervision. A copy of their memo covering procedures and date of count must be furnished to the Internal Auditing department, who may field in a representative to observe the count.

The Warehouse Head must furnish the project accountant with copy of Packing and Shipping Lists covering acquisitions of tools. The letter is responsible for recording the accountability of the tool room in charge and to make such verification as are necessary, periodically.

9.4   TOOL ROOM

Tool rooms must be constructed integral to the shop area and a standard warehouse control must be applied to control the tools borrowing and return. This is to avoid theft and pilferages and must be constructed with adequate protection against unauthorized withdrawal. Permanent personnel can be assigned as tool room keeper or in charge.

Adequate space for the tool room must be provided. A typical area should be divided in sections of office desk and files, mechanics or automotive tools, hand tools, power tools and accessories (drilling bits, blades, pads and grinders), small compartments for chain hooks, links, shackles, turnbuckle, clamps and binders and claws.

An area or open space yard for wire ropes and various sling assemblies must be included. All tools and accessories should have been properly issued by the warehouse and accountable to the tool keeper.

Inventory should be done every year, and equipment in disrepair must be reported and replaced.

Part IV

Project Operations

10 Equipment Project Schedule

Equipment is one of the main components of the construction project. As such, equipment is just a crucial and critical resource to deliver on time or on schedule their assignment. A complete equipment schedule and operating procedure are paramount to this effect. The equipment must have a procedure from project start to finish, which must be the same and consistent for all projects a piece of equipment is assigned. The underlying procedure is tested and can ensure smooth and problem-free equipment operation. This will allow equipment to be transferred or moved again and again to different projects but never lose track of maintenance, operation and equipment cost and time assessment.

The master project schedule is the detailed list of all construction activities, from the start of the project up to the finish time or completion of the project. From this, a second schedule is prepared to detail the durations and milestones of each activity and the resource and establishes deliverables of each activity. A third project schedule is prepared to only include the project activities and the heavy equipment requirements of each activity (see Attachment 4).

As this is a time schedule, equipment managers will assign the specific heavy equipment to use for each project phase up to the end of all activities. For example, for excavation, the earthmoving equipment will be assigned, such as excavators, bulldozers and dump trucks.

Effective project equipment schedule is a critical component of project planning as it is time management of equipment assignment for each phase of the project. The most popular form of project equipment schedule is the Gantt chart.

Benefits of project equipment scheduling in project management:

- Determines the equipment resources and quantity needed by the project. In case of non-availability of equipment, the option to rent will be discussed.
- Establishes the duration of each piece of equipment assigned to the project. This will be used to anticipate maintenance scheduling of the assigned equipment.
- Becomes the basis of field equipment's office composition based on the quantity of equipment, up to project completion. This will set up facilities and staff for the tracking, monitoring, reporting, maintenance and operation of the equipment in the field.

The equipment project management must have a master equipment schedule complete with the equipment number, quantity, project deployment duration, implements and accessories required. The duration must be in a number of months to days to be used by the project. If the project is equipment intensive, this is even more important to have at the beginning of the project. The equipment schedule must be prepared by the Technical Services and Planning section and submitted for approval by Equipment Head, then submitted for approval of the assigned Project Manager who shall integrate it into his final project schedule.

Project equipment scheduling is intended to match the resources of equipment, materials and labor with the project master schedule. The equipment field operations in return will prepare the project, preventive maintenance schedule, multiple movement operations and materials and accessories needed for operations and maintenance of the equipment fleet for and up to the total project duration and completion. Additionally, the schedule will also be a big factor to manage equipment productivity, cost and other reports such as time and cost of the equipment while assigned to the project

11 Project Operating Forms and Instructions

11.1 EQUIPMENT INSPECTION CHECKLIST

To be presented here are two types of equipment as models of inspection checklist of equipment. These are the vehicular and earthmoving equipment checklists that are maintenance forms that guide equipment inspectors during technical inspection of equipment.

The forms indicate which parts per grouping shall be checked wherein corresponding conditions are reflected. The information gathered would indicate to maintenance services the extent and type of repairs most essential for the equipment.

Equipment inspection shall be conducted under any of the following transactions:

1. During receipt of new acquired equipment.
2. Before shipment of equipment to and from project.
3. Before and after major equipment repairs.
4. During execution of PM 1–3, 1–4 and 1–5.
5. At any time equipment warrants inspection.

11.1.1 INSTRUCTIONS ON INSPECTION PROCEDURE

Inspect the equipment in sequence as indicated in the checklist. Opposite to each item, indicate annotations accordingly. Remarks on notations shall be the same as the legend, i.e. "OK" for satisfactory or complete set up, "R" for repair, "PM" for service and "MP" for missing parts.

On completion of inspection, show other defects found on items that are not included in the checklist. The remarks and recommendation column shall show informative comments that are helpful for maintenance services, e.g. what repairs are necessary, what preventive measures to avoid recurrence of defects noted, what changes if there are needed changes, all must be specific and clear.

11.1.2 RESPONSIBILITY TO INSPECT EQUIPMENT

Primary responsibility of inspection shall be on PM section of the Technical Services, which shall have an assigned Equipment Inspector to inspect and test all incoming and outgoing equipment at the central shop and yard. The Inspector must possibly also be qualified as maintenance mechanic and operator/driver of equipment.

They can evaluate and recommend corrective action on the noted defects and other findings. After approval and notation on the result of inspection, copy shall be furnished to the shop head for any action needed on the findings if any.

Company Name & Address   Form No. _____

Inspection Checklist for Crane Equipment   Date: _____

This form indicates the scope of Technical Inspection for EARTHMOVING EQUIPMENT preparatory to issuance and receipt to and from projects.

EQUIPMENT NO: _____   DATE _____

DESCRIPTION: _____   HOURMETER/ODOMETER Reading: _____

Reason for Inspection: Please check: _____Incoming _____ Outgoing _____ PM

LEGENDS:	Satisfactory	-	OK	Service	-	PM
	Adjust	-	Ad	Missing	-	MP
	Repair	-	R	Replaced		RP

1. ENGINE ACCESSORIES

_____	Cylinder Head & Gasket	_____	Vacuator Valve
_____	Cylinder Head & Gasket	_____	Vacuator Valve
_____	Radiator Assy.	_____	Turbo Charger
_____	Radiator Cap.	_____	Blower Assy.
_____	Radiator Hose/Clamps	_____	Fuel Tank Cap.
_____	Thermostat	_____	Fuel Lines/Hoses/ Connections
_____	Water Pumps	_____	Fuel Filter Housing
_____	Fan Belt	_____	Carburator/Injection Pump
_____	Oil Filter Assy.	_____	Sparkplug/Injectors
_____	Oil Cooler	_____	Fuel Pump/Seals/Gasket
_____	Water Filter	_____	Battery Housing
_____	Fan Belt	_____	Carburator/Injection Pump
_____	Oil Filter Assy.	_____	Sparkplug/Injectors
_____	Oil Cooler	_____	Fuel Pump/Seals/Gasket
_____	Water Filter	_____	Battery Housing
_____	Air Cleanser Assy.	_____	Battery Post
_____	Air Cleanser Element	_____	Battery Clamps/Terminals
_____	Battery Cells/Caps/	_____	Starter Assy
_____	Electrolyte	_____	Voltage Regulator
		_____	Engine Regulator

2. STARTING ENGINE & ACCESSORIES

_____	Air Cleaner
_____	Cylinder Head, Manifold
_____	Crankcase Breather
_____	Ignition System-Magneto/Distributor
_____	Sparkplugs-coils
_____	Fuel Pump/Carburator/Governor

3. Instrumentation, Lights & Control (cont)

_____	Blade Tilt Control (L/R Ends)
_____	Circle Side Shift Control
_____	Circle Reverse Control
_____	Wheel lean Control
_____	Ripper Control
_____	Scarifier Control Lever

_____ Clutch Transmissions
_____ Heater Plug
_____ Fuel Tank Lines

3. INSTRUMENTS, LIGHT & CONTROLS
_____ Switch Ignition
_____ Switch Starter

_____ Headlight Guard
_____ Tail & Stop Light
_____ Signal Light & Switch
_____ Governor Control Level
_____ Steering Wheel Lever

4. UNDERCARRIAGE & BRAKE GROUP
_____ Tie Rod Ends, Pitman Arms
_____ Brake Drum, Dust Shield
_____ Axle Assy.
_____ Axle Shaft
_____ Propeller Shaft
_____ Spring Suspensions
_____ Air Tank Lines
_____ Seal & Oil Retainer (Inner & Outer)
_____ Wheel Alignment
_____ Tires & Rims
_____ Track Assy. (Shoe & Pads)
_____ Track Carrier/Track Rollers
_____ Track Adjuster
_____ Sprocket, Bolts, Drive Chains
_____ Track Idlers
_____ Frame & Guards
_____ Spring &Pads
_____ Track Tension/Adjustment
_____ Steering or Travel Clutch
_____ Final Drive Housing, Seals & Gaskets
_____ Propeller Shaft U-Joint Bolts
_____ Propeller Universal & Slip Joints
_____ Dump Cylinder
_____ Tilt Cylinder
_____ Strut Assy.
_____ Twins Hits Assy. & Control

6. CRANE UNIT & ATTACHMENTS
_____ Turntable Assy.
_____ Swing Control Lever & Treadle
_____ Hoist Mechanism
_____ Hoist Clutch

_____ Parking Brake Control Lever
_____ Bowl Control
_____ Apron Control

_____ Boom Control
_____ Speed Selector Safety Lock Control
_____ Krugger Moment & Anti-block System
_____ Service Motor
_____ Converter Temperature
_____ Blade Tilt Control Lever
_____ Clutch Pedal Lever

5. ATTACHMENTS
_____ Ripper Assy./Cylinder
_____ Ripper Tips
_____ Cutting Edges
_____ End Bit Teeth
_____ Molboard Pins/Pivot Socket, Bolts
_____ Circuit Drive Housing
_____ Side Arms-Mounting
_____ Circle Reverse Drive Housing
_____ Blade Lift Drive Housing
_____ Circle Center Drive Housing
_____ Circle Drawbar
_____ Front Axle & Wheel Steering
_____ Center Shift Link
_____ Cantiler Yoke Assy.
_____ Push Beam
_____ Springs & Stop Blocks
_____ Hydraulic Motors
_____ Flight Conveyors
_____ Ejector Cylinder
_____ Apron
_____ Routing Bit
_____ Draft Frame Side Arm
_____ Boom Stopper
_____ Tagline Cables
_____ Fairlead/Suspension Cable
_____ Boom Section Control Lever
_____ Swing Brake

_____ Crane Lever Selector
_____ Hoist Speed Selector
_____ Main Hoist Control Lever
_____ Swing Lock Operations

_____	Hoist Drum	_____	Telescopic Control
_____	Disconnecting Lever Clutch		
_____	Power Boom (Lowering & Derricking)		
_____	Boom Attachments	**7. HYDRAULIC SYSTEM**	
_____	Crane Hook & Block	_____	Hydraulic Hoses & Fittings
_____	Cables	_____	Hydraulic Pump & Tank
_____	Sheaves Pulley	_____	Breather
_____	Outriggers Control	_____	Control Valves
_____	Hoist Assy.	_____	Hydraulic Cylinders
_____	Swing Assy. Gears Circle		
_____	Roller Path/Roller		
_____	Boom Indicator		

8. LUBRICATION

_____ Engine Crankcase Oil

_____ Trans. Oil

_____ Differential Oil

_____ Steering Gear Housing Oil

_____ Torque Converter Oil

_____ Hydraulic System

_____ Drive Case Oil

_____ Final Drive Tandem Housing

9. MISCELLANEOUS

_____ Fire Extinguisher

_____ Road Permit

_____ Insurance Coverage

_____ Manual (Parts, Service, etc.)

_____ PM Folders

OTHER DEFECTS NOTED: _____

_____

REMARKS AND RECOMMENDATIONS:

_____

_____

_____

INSPECTED BY: CERTIFIED BY:

_____   _____
Equipment Inspector   Equipment Supervisor/Superintendent

_____
Mechanic/Electrician

Company Name & Address   Form No._____

Inspection Checklist for Vehicular Equipment   Date:_____

This form indicates the scope of Technical Inspection for Vehicular Equipment preparatory to issuance and receipt of equipment to and from projects.

EQUIPMENT NO: _____   DATE: _____

DESCRIPTION: _____   HOURMETER/ODOMETER Reading: _____

Reason for Inspection: Please Check: _____ INCOMING _____ OUTGOING _____ PM

LEGENDS:	Satisfactory	-	OK	Service	-	PM	Not Applicable - NA
	Adjust	-	AD	Missing	-	MP	
	Repair	-	R	Replaced	-	RP	

I. ENGINE & ACCESSORIES

_____ Cylinder Head/Gasket	_____ Generator/Alternator
_____ Engine Block	_____ Battery Box
_____ Crankcase/Gasket	_____ Battery Cable, Clamps, Terminals
_____ Radiator Cap	_____ Battery Caps
_____ Radiator Hoses	_____ Battery Posts
_____ Water Pump	_____ Battery Solution Level
_____ Thermostat	_____ Distributor Assy.
_____ Air Cleaner Assy./Vacuator Valve	
_____ Air Hoses/Clamps	
_____ Restriction Indicator	
_____ Fuel Tank	
_____ Fuel Filter Housing	
_____ Fuel Filter	
_____ Fuel Lines/Fittings/Connections	
_____ Carburetor/Injection Pump	
_____ Sparkplug/Injectors	
_____ Oil Filter Assy.	
_____ Starter Assy.	

2. CLUTCH, TRANSMISSION & TRANSFER CASE

_____ Clutch Pedal	_____ Master Cylinder Assy.
_____ Linkage/Bearing Release.	_____ Hydrovac Assy.
_____ Clutch Lining	_____ Fluid Lines & Flexible Hoses
_____ Shift Lever, Main Transmission	_____ Brake Lining/Drum
_____ Shift Lever, Aux. Transmission	_____ Air Tank
_____ PTO	_____ Air Lines & Brake Chamber
_____ Transfer Case Assy.	_____ Hand brake Assy.
_____ Main Transmission Assy.	_____ Slack Adjuster
_____ Aux. Transmission Assy.	

3. DIFFERENTIAL ASSY. & UNDERCHASSIS

_____ Front Differential & Axle Assy.	_____ Torque Red Assy.
_____ Intermediate & Axle Assy.	_____ Chassis Assy.
_____ Rear Differential & Axle Assembly	_____ Cross Members
_____ Propeller Shafts	_____ Wheel Alignment
_____ Cross Joints	_____ Tires & Rims
_____ Shock Absorber Assy.	_____ Stud Bolts & Nuts
_____ Leaf Spring Assembly	

4. STEERING & BRAKES

_____ Steering Wheel
_____ Draglinks
_____ Pitman Arms
_____ Tie Rod Ends
_____ Brake Pedal
_____ Master Cylinder Assy.
_____ Hydrovac Assy.
_____ Fluid Lines & Flexible Hose
_____ Brake Linings/Drums
_____ Air Tank
_____ Air Lines & Brake
_____ Air Lines & Brake Chamber
_____ Hand Brake Assy.
_____ Slack Adjuster

5. CAB & BODY

_____ Deck Top Assy.
_____ Doors, Hinges Latches
_____ Seats & Backrest
_____ Rear View Mirror
_____ Windshield
_____ Wiper
_____ Body

6. ELECTRICAL & INSTRUMENT PANEL

_____ Switch, Starter
_____ Switch, Ignition
_____ Headlight Assy. & Switch
_____ Signal Light & Switch
_____ Park Light & Switch
_____ Horn Assy. (Air/Electrical)
_____ Fog Light & Switch
_____ Ammeter
_____ Hour meter
_____ Speedometer
_____ Odometer
_____ Air Pressure Gage Assy.
_____ Fuel Gage Assy.
_____ Tachometer Assy.
_____ Tachograph Assy.
_____ Mixer Control Unit

7. SPECIAL ATTACHMENT

_____ Dump Body Assy.
_____ Mixer Assy.
_____ Hydraulic System
_____ Hydraulic Pump Assy.
_____ Fifth Wheel Assy.
_____ Power Shift/Torque Converter Assy.
_____ Hydraulic Wrecker Assy.
_____ Distributor Burners
_____ Tank Mounting

8. LUBRICATION

_____ Engine Oil
_____ Main Transmission Oil
_____ Aux. Transmission Oil
_____ Front Differential
_____ Intermediate Differential
_____ Rear Differential Oil
_____ Hydraulic Oil
_____ Torque Converter Oil
_____ Lower Steering Oil

9. MISCELLANEOUS

_____ Spare Wheel & Tires
_____ Emergency Tools (Tire Wrench, Jacks, etc.)
_____ Manuals (Parts, Service etc.)
_____ PM Folders, Time & Cost, etc.

OTHER DEFECTS NOTED: _____

_____

_____

REMARKS & RECOMMENDATIONS: _____

_____

_____

_____

Equipment Inspector   Equipment Supervisor/Superintendent

_____

Mechanic/Electrician

11.2   EQUIPMENT TIME AND COST

The Time and Cost Report is the record of the summation of equipment time, cost, maintenance, both PM and corrective, breakdown or idle period, POL consumption and manpower cost. Data is taken daily while an equipment is assigned at the project site. The daily equipment time and operating costs whether for operation or maintenance purposes are its primary input. See attached Time and Cost Form (see Attachments 1 and 2).

The Annual Time and Cost Summary Report is a translation of the Monthly Time and Cost Report. Both are found in a folder and are transmitted with the equipment wherever it goes. Care must be taken whether it is in a sealed plastic envelope so as not to get wet when there is rain or too much dust from sandstorms. The equipment must be kept and filed up by the PM clerk or PM engineer at the project site.

Upon return of the equipment to the central yard or to another project assignment, the folder must likewise be sent. The receiving end must check it as among items with the equipment.

The Summary Sheet and the Monthly Time and Cost Report reflect the following information.

1. The total and a breakdown of the equipment time.
2. The total and a breakdown of the equipment repair costs.
3. The total and a breakdown of fuels and lubricants issues and costs whether for operation or maintenance use.
4. The total filter costs.
5. The total and a breakdown of operator's wages.
6. The average hourly operating cost.
7. The average fuel rate.
8. The average percentage of (%) utilization
9. The average percentage of (%) availability

Content details are as follows:

1. Equipment time:
   1.1   Available – equipment mechanically operable, sub-divided into:
         a.   Operating
         b.   Idle
   1.2   Breakdown – equipment not mechanically operable, sub-divided into:
         a.   Undergoing repairs
         b.   Not undergoing repairs
         c.   Undergoing PM services
   1.3   Total shift hours – the sum of available and breakdown hours.
   1.4   Utilization hours – the length of time the equipment is engaged or working at the project site.
2. Remarks – reasons why equipment is not-in-production, e.g.:
   a.   Breakdown – state nature of breakdown.
   b.   Idle due to unfavorable weather and/or unfavorable working condition.
   c.   Idle hours of this nature shall not be included in the computation of percent (%) utilization
   d.   Other incidents, accidents or delays.
3. Parts cost – the total cost of spare parts, materials and supplies expended in the repair. The total cost is enough for repair taking more than one day to finish instead of making daily entries.
4. Man-hours – the total labor hours expended in the repair. The total man-hours is enough for repairs taking more than one to finish instead of making daily entries.
5. Labor cost – the corresponding cost of man-hours.
6. Fuels – either gasoline or diesel; indicate quantity issued and cost. Also indicate whether issue is for operation or maintenance purposes.
7. Lubricants – indicate the type, quantity and cost of lubricant issued and whether for operation or maintenance use.
8. Filters – indicate the type, quantity and cost.
9. Operator's time and wage.
10. SMR – Service Meter Reading.
11. Type of work done – this is the work performed by the equipment, i.e. dozing for bulldozers and loading for loaders.

11.2.1   Summary Data (below the form)

1. Operating cost for month – it is the total cost of parts, daily maintenance for the month.
2. Meter reading – indicates service meter reading at the start and at the end of the month. The difference should be approximately equal to the total operating hours.

Computed data for the following:

1. Hourly operating cost = Operating cost for the month/total operating hours
2. Fuel rate = Total monthly issue/total operating hours
3. Percent utilization = Total operating hours/total available hours × 100%
4. Percent availability = Total available hours/total shift Hours × 100%

11.2.2 Notes on Availability and Utilization Reports

With the utilization report, the equipment team will see exactly how much equipment time, as scheduled, is employed and utilized in the project. Using equipment means the resources being utilized are available at the right scheduled time for the right activity for the right project assignment.

In essence, it is the process of successfully distributing equipment resources to meet the objectives of the project.

With the availability report, the overall assignments of equipment can be planned and prioritized. The project can know which equipment shall remain engaged in the project up to the completion of each phase of work. Project management can then have the option of assigning equipment according to the project phase of work, period of use and duration in each assignment. This will give the flexibility to project management on how to assign equipment more effectively and efficiently. Project duration, cost of transport and priorities will come forth for decisions as to how the equipment for which projects will be assigned.

Other important information on availability and utilization will be the option to outside lease equipment to fill up the work of each piece of equipment in the project.

11.2.3 Route of the Report from the PM Engineer at the Project

Requisitioning and inventory of forms	PM Engr.
Distribution to Project Equipment Clerk	PM Engr.
Filling up to form	Project Equipment Clerk
Confirmation of entries	PM Engr.
Certification of accuracy of entries	Equipment Supt.
Transmission of Central Office	PM Engr.
Review of reports	PM central office

To computerize the Time and Cost Reports, use Microsoft Office: Excel program.

1. Preparing a Microsoft Workbook using the Time and Cost forms (see Attachments 1 and 2).
2. Assign and prepare one Excel Workbook for each equipment type, e.g. WL (Wheel Loader), DT (Dump Truck) and for all the 40 or more equipment-coded units (see Table 4.1 in Chapter 4, Part II of this book).

3. In each Excel Workbook, assign or name one sheet per month to make the January–December months or sheets 1–12. For the 12 sheets, use the Time and Cost Monthly Form, see Attachment 1. Assign the last sheet for the Time and Cost Yearly report form and use the year-end summary report form (see Attachment 2).

4. Microsoft Excel has its own mathematical formulas to add and total each column, horizontally, for the totals and use the given multiplication and division formulas, in the form to get the percentages (see the Attachments).

Other forms and reports for daily gathering, accumulating and transposing of pertinent data to complete entry for the Time and Cost Sheet (Table 11.1).

End of the month (fill in accordingly)

1. Monthly totals
2. Motor reading – end of month
3. Operating cost for month
4. Hourly operating cost
5. Fuel rate
6. Compute percent (%) utilization
7. Transposing monthly totals to annual summary sheet
8. Affixing signature
9. Submitting to PM engineer

End of the year (fill in accordingly)

1. Yearly totals
2. Operating cost
3. Hourly operating cost
4. Fuel rate
5. Percent (%) utilization
6. Percent (%) availability

The report must adhere to some few important steps:

1. Reviewing and evaluating newly received record booklet from other project or central yard. Specially noting down previous repairs done to serve as basis for future PM plan and action.

TABLE 11.1

Type of Forms	Source	Supervisor
DETC for equipment time	Operator/Driver	Superintendent/Truck master
Job Order form for man-hours and labor cost	Shop Mechanic/Elect/Technician	Shop Superintendent.
PM reports	PM Engineer/Clerk	PM Engineer
Parts costs	Warehouse man	Warehouse Head
Operator's time and wage	Accounting/Personnel	Timekeeper

2. Emphasizing the rule that entries shall be made daily.
3. Accumulating the forms at the end of each month for review and correction purposes.
4. Submitting forms to the Equipment Superintendent for review.
5. Submitting duplicate copies of forms to central office so as to be received on or before 5th of each ensuing month.

11.2.4 IMPORTANT NOTES FOR THIS FORM PROCEDURE

1. Operational use – fuels and lubricants withdrawn are for operation purposes or for the accomplishment of its work or function.
2. Maintenance use – fuels and lubricants withdrawn are for maintenance proposes, i.e., for washing, break-in, test runs, repairs, trouble shooting and PM services.
3. Project copies of 1 year record booklet must be kept and maintained properly and shall always be with the equipment in case of transfer.
4. The number of Time and Cost Record Booklets that shall be with the equipment at all times shall be equal to the equipment age, i.e., if the equipment was acquired 3 years ago, there must be three booklets in the equipment jacket or folder.
5. Equipment made up of composite units shall be prepared with separate Time and Cost Sheet and booklets, i.e., truck-mounted cranes compose of Crane and Truck units and Asphalt Distributor composed of Distributor and Truck Units Plants.
6. Only major equipment shall be prepared with Time and Cost Sheet and booklet. If the company has process plants, only one booklet shall be made per one process plant.

11.3 EQUIPMENT FAILURE REPORT

The Equipment Failure Report is a project maintenance form to document the failure on any part of the equipment during operation. It shall provide maintenance personnel and services informative observation and comments that caused the failure during operation or while at the job site.

The report can also be used for failure observed during the first inspection upon delivery of the equipment. The Equipment Inspection Checklist form maybe use for more multipart failure of equipment.

Equipment Failure Report shall be prepared under any or all of the following circumstances:

1. When failure occurs within the warranty period.
2. When failure appears to be due to improper manufacture or assembly.
3. When failure appears to be due to improper or inadequate lubricant, or if lubrication service is insufficient.
4. When failure indicates that failed part is of inferior material, poor quality and inadequate for its function.

5. When failure appears to be due to negligence or insufficient technical know-how of responsible personnel such as follows:

   a. Failure of operator to report immediately any observable defects or suspected faulty operation of his assigned equipment.

   b. Failure of operator to properly operate the equipment especially when operating in a very unstable condition in the project area.

   c. Failure of personnel responsible for equipment repair to correct all defects noted.

   d. Failure due to improper or poor quality workmanship or repairs done to equipment or components.

   e. Failure of those responsible on equipment PM to implement the lubrication and general care of equipment based on manufacturer's recommendation.

   f. Failure due to minor defect that was taken for granted such as loose or missing hardware, inoperative gauges, cracks, oil leaks, missing radiator cap, chaff hoses and wires and underinflated tires.

Typical Failure Report Form

	Form No._____
Company Name and Address Equipment Failure Report	Date: _____

Equipment No.	Date Put in Operation
Description:	Date of Failure
Make& Model	HRS. OR KM. To Failure
Chassis Ser. No.	
Engine Serial No.	Last Operator Name:
Serial No. (Other):	Last Operator ID Code:
NATURE OF FAILURE(indicate Assembly/Part affected):	FAILURE ANALYSIS:

RECOMMENDED CORRECTIVE ACTION:

Submitted By: Equipment Supervisor	Checked By: Equipment Superintendent	Noted By: Project Manager

11.4 EQUIPMENT ACCIDENT REPORT

Project site all differ from each other from one country to another country. Both projects maybe for dam construction, but the environment and earth configuration would differ in many various ways. Each project should identify these possibilities at the start of the project.

In case of accident, a report must be made by the supervisor or superintendent. It should contain the details of the probable cause of the accident. A typical report form for accident is shown below.

Typical Accident Report Form

Form No._____

Company Name & Address
Equipment Accident Report

| Equipment No: _____
Make & Model: _____ | Project Name: _____ | Date: |
| Operator Name: _____
ID Code: _____ | Project Location:_____ | Time: |

Equipment Accident Description and Cause:

Affected Equipment Assembly and/or Parts:

No.	Part Name & Number	Damage Description	Status: (for repair or for junk)	Remarks

Overall Equipment Assessment (Keep at project site for repair, or ship back to central shop for major repairs);

| Prepared by:
PM Engineer | Noted by:
Equipment Superintendent: |

12 Project Maintenance Standard Procedure

12.1 EQUIPMENT PM SCHEDULING

The monthly PM schedule report is a maintenance form to document the forecasted schedule of equipment at the project site for preventive maintenance services for one-month period (see Attachment 3).

The form is designed to enable project maintenance services to plan the manpower and materials requirement for preventive maintenance purposes at the job site.

PM number and operating hours (see Maintenance) due shall be scheduled in the standard intervals as follows:

PM 1	-	60 Hours	PM 1	-	560 Hours
PM 1-2	-	125 Hours	PM 1-2	-	625 Hours
Pm 1	-	185 Hours	PM 1	-	685 Hours
PM 1-3	-	250 Hours	PM 1-3	-	750 Hours
PM 1	-	315 Hours	PM 1	-	815 Hours
PM 1-2	-	375 Hours	PM 1-2	-	875 Hours
PM 1	-	435 Hours	PM 1	-	935 Hours
PM 1-4	-	500 Hours	PM 1-5	-	1000 Hours

12.1.1 PROCEDURE

1. Enter the equipment number in the space provided for.

   1. Based on the last PM service of previous month, enter the first PM schedule for the month on the dates column. The upper column shall show PM number and operating hours due in the lower column. The succeeding PM schedule shall be forecasted by dividing the PM interval in days. This shall be the number of days before the next PM schedule. This process is repeated for other units.
   2. The remarks column shall show informative comments that are helpful for maintenance management, e.g., reasons why PM is not scheduled or performed.

12.1.2 RESPONSIBILITIES

The primary responsibility is delegated to the PM Engineer at the project and to whom PM scheduling is assigned. The main administrative responsibility is delegated to the Technical Services Head.

DOI: 10.1201/9781003360667-16

As most of the equipment should be at the project site, the equipment head at the project site shall foresee the schedule preparation to ensure the schedule is prepared and followed at the project site.

During the period the equipment is received at the job site, the following are immediately checked and performed.

1. Check the previous PM services of the equipment as given in the time and cost record of the equipment.
2. Review, evaluate and initiate correction action if any, to update the PM services as reported in the time and cost record.
3. Prepare the PM schedule for the equipment as per the given form below.
4. Establish and approve schedule of PM services for the next subsequent one-month period as necessary.
5. Enter the schedule in the PM Schedule Board of the project.
6. Enforcing to all levels of responsibility for the strict compliance to the PM schedule and instruction.

Sample of Monthly PM Schedule (Projected)

Company Name & Address

Monthly PM Schedule (Projected)

Form No. ———

MONTH ——————, Date —————— Year ——————

Project Name: ——————
Location: ——————

Equip. No	1	2	3	4	5	Same columns from 6 to 24....	25	26	27	28	29	30	31	REMARKS
CT - 08														
TT - 22														
MG - 12														
CT – 18														
DT – 25														
DT - 37														
GP 112														
SV - 59														
SV – 51														
LP - 44														

Prepared By: ——————
PM Engineer: —————— Equipment Superintendent: ——————

12.2   PREVENTIVE MAINTENANCE SCHEDULE BOARD

All projects and job sites must have and display a Preventive Maintenance Schedule Board at the shop office.

The PM schedule board shall be a ¼" × 4" × 8" plywood with green paint background and white lines and letters.

Responsibility in accomplishing the board is delegated to the project PM Engineer or to any person to whom PM responsibility is assigned in case the job site is small.

Central shop and yard shall likewise display the same schedule board for that equipment utilized at the shop and for the maintenance of the yard.

◄─────────────── 4' ───────────────►

PM SCHEDULE BOARD							
EQUIPT NO.	PERFORMED			DUE			REMARKS
	PM NO.	OPHRS	DATE	PM NO.	OPHRS	DATE	
6"	5"	5"	5"	5"	5"	5"	12"

12.3   LOBE OIL OPTIMIZATION PROGRAM

The lobe oil optimization drain period shall define the extension of oil drain interval for the engine given as examples below. The extended drain period is the result of monitoring used oil condition from some 30 various applications and operating conditions.

Typical drain intervals (For Caltex RPM DELO 400 Oil SAE Only):

1. Detroit Diesel Engine (all models) – extended drain interval from 125 hours to 250 hours.
2. Caterpillar Engines (all models) – extended drain interval from 250 hours to 315 hours.
3. Dump Truck (12 cu yard) Engine – extended drain interval from 4,000 km to 7,500 km or approximately 250 hours to 450 hours.

These intervals can be used on the first PM schedule that shall coincide with new drain interval.

During the monitoring period, 15.2% and 7.6% of the sample tested indicated fuel dilution and excessive insoluble, respectively. The rest are suitable for further use. Analyses indicated these were caused by faulty injectors, low compression, etc. and not the result of poor oil quality. The sample oil, after correction shall be made on engines, indicates again suitability for further use.

In essence, in order to enjoy the full advantage of an extended drain that comes along with the use of high additive content oil, the exercise of a diligent PM program must be followed.

A viscosity Comparator Tester will provide quick means of detecting abnormalities before the drain period for the project sites.

13 Project Tower Crane Inspection

A construction tower crane is a common lifting equipment used for high-rise construction projects. It is called a tower crane because of the large tower-like structure the crane is mounted and pivots around. The tower mast supports the jib and counter jib which in turn supports the cabling, trolley and hook which does the lifting.

Traditional mobile cranes require large working space. While the reach of the tower crane (jib and hook locations) is large and long, the actual tower crane space set-up takes up on-site is relatively small and the boom reach can be long and rotate around 180° with little obstruction.

To ensure the utmost safety, it is important to perform an inspection check of the tower crane before operation. This is due to the high risk involved in operating this equipment.

13.1 THINGS TO CONSIDER WHEN DECIDING ON A TOWER CRANE

If you're planning a tower crane on your project site, it's important to take a number of key factors into consideration. The key considerations to take into account when planning for a tower crane are discussed below.

13.1.1 CAPACITY REQUIRED

For lifting of formworks and buckets of concrete, use of crane will be a relatively light model. For lifting of equipment such as a generator or large steel members use you will consider crane for critical lift. Cranes have typical LOAD CHARTS that are intended for all loads especially the critical lift. Critical lifts are typically those that are within 10% of the crane's total capacity. These lifts require special engineering. If you only have one heavy lift on the project, it may be worthwhile to engineer your crane to accommodate all other loads except the one and have it engineered as a critical lift. Your other alternative could be to bring in a mobile crane to complete the lift.

13.1.2 SPACE AVAILABILITY

It is critical when planning your tower crane location that there are no obstructions and that the crane is clear to turn around its operating radius.

DOI: 10.1201/9781003360667-17

In instancing where more than one tower crane is on-site, the swing radius of each crane should be considered, optimizing the area of each lift to provide the most coverage while not hitting the other cranes' towers while rotating.

13.1.3 PRE-PICK DOCKS LOCATIONS

Planning your pick locations around your tower crane OR planning your tower crane around your pick locations is important. Loading docks are used as material come into and out of your job site. Cranes get weaker the further the load is from the mast – plan your crane accessibility to these locations to maximize the capacity of your crane.

13.1.4 FLIGHT PATHS

If your crane is located within a certain distance to an airport, it may require a special permit and restrictions on its operating time/height. Review with your local aviation authority to ensure that you take out the proper paper work prior to erecting your crane.

13.1.5 LOAD RATING

Every tower crane should be equipped with a substantial and durable LOAD CHART, with clearly legible letters and figures and having the following information:

1. A full and complete range of the manufacturer's approved crane load ratings of all stated operating radii (or jib angles) for each recommended counterweight, jib length, tower height or other installation conditions.
2. Lifting speed instructions; operators must be advised on the extent and limits of the crane speed for all loads and lifts.
3. Recommended parts of hoist reeving, size and type of rope for various crane loads.
4. Essential precautionary or warning notes relative to limitations on equipment and operating procedures.
5. Drum data, available line pull, permissible line pull, line speed and rope spooling capacity.
6. Wind velocity operating limits. Note the changing wind directions and the weather periods that come with changing wind velocity. Must be in the crane plan.
7. LOAD CHART should be securely attached to the cab in a location easily visible to the operator while seated at his control station.
8. When crane is operational from the remote control console, the LOAD CHART should be attached to a substantial plate secured to the console.

13.2 PRECAUTIONS FOR MORE THAN ONE CRANE OPERATING IN THE SAME AREA

Minimize the probability of collision on cranes or the hoisting ropes or loads fouling each other by observing the following precautions:

1. Prepare a crane set-up plan before erection of the cranes.
2. The crane should be located in such a way that the operators have a clear view of other cranes operating within collision danger areas.
3. The operators should have direct communication with each other so that one operator may alert the other of impending danger.
4. Overall lifting program should be set out and controlled by an engineer who is in contact with all operators and slingers and in charge of assigning priority of operation for all the cranes.

13.3 TOWER CRANE PREVENTIVE MAINTENANCE CHECKLIST FORM

Tower Crane Inspection is very important step for tower cranes. In order to retain the safety of tower crane, tower crane inspection and maintenance should be carried out at least once in a month by the inspection and maintenance technicians for tower crane. Basic purpose of the inspection is to reduce the risk. The inspection technicians should properly record all the details as given and their findings. The record should be read and signed by inspection and maintenance technician(s) (Figures 13.1 and 13.2).

FIGURE 13.1 Two tower cranes in operation.

FIGURE 13.2 Multi tower crane operation.

TOWER CRANE PREVENTIVE MAINTENANCE CHECKLIST FORM

Company Name: _____   Date: _____

Equipment No. _____   Description: _____

Inspected by _____   Project: _____

PREVENTIVE MAINTENANCE CHECKLIST (PAGE 1 OF 3)

1.0 General:
- body metal panel
- guards & cover panels
- external lights
- safety / warning decals & labels
- **hand signal chart**

2.0 Driver's Cab and Station:
- grab rails & steps
- glass
- windshield wipers
- door restraints
- mirrors
- fire extinguisher
- seat restraint(s)
- seat belt(s)
- parking brake
- air pressure
- instruments & gauges
- switches
- horn
- lights
- steering
- engine clutch
- accelerator
- brake

5.0 Carrier Power Plant (lower):
- exhaust system guards & insulators
- belts & hoses
- guards/covers

6.0 Carrier:
- transmission
- drive line
- tires/wheels
- main frame members
- hydraulic hoses, tubing & fittings

- hydraulic fluid level
- anti-skid surface
- axle lock-out
- backup alarm and light

3.0 Outriggers:
- boxes
- beams
- cylinders
- floats/pads
- hydraulic hoses, tubes & fittings
- holding valves
- position locks
- warning signs

4.0 Operator's Cab & Station:
- grab rails/steps/platforms
- anti-skid surfaces
- glass
- windshield wipers
- door restraint
- fire extinguisher
- mirrors
- seat restraint
- seat belt
- operator's manual
- operating instructions/decals
- electrocution warning sign (inside)
- hand signal chart
- parking brake
- swing brake
- positive swing lock
- controls – forces/movements
- accelerator/throttle control
- air pressure
- hydraulics
- horn/warning device

PREVENTIVE MAINTENANCE CHECKLIST (PAGE 2 OF 3)

7.0 Load Chart:
- per configuration
- durable, legible, visible from Operator's station
- secured
- Safety Devices/Operational Aids:
- boom angle indicator
- boom length indicator
- main drum rotation indicator
- auxiliary drum rotation indicator
- load moment indicator
- load weight indicator
- radius indicator
- crane level indicator
- anti-two block device
- two block warning / damage device

9.0 Rotating Upper Structure:
- turntable
- electrical collector ring
- counterweight frame
- hydraulic pump(s)
- hydraulic hoses/tubes/fittings
- hydraulic pressure
- electrical wiring
- main hoist – motor/valves/lines
- main hoist – wrapping on drum
- main hoist – minimum (2) rope wraps
- auxiliary hoist – motor/valves & lines
- auxiliary hoist – wrapping on drum
- auxiliary hoist – minimum (2) rope wraps
- counterweight/mounting
- swing gear box
- electrocution warning sign (outside)
- counterweight warning sign

8.0 Main Boom:
- lift cylinder(s)
- telescoping cylinder(s)

- hydraulic hoses/tubing & fittings
- holding device
- boom sections alignment
- wear pads
- equal extension
- sheaves
- hoist line dead end
- wire rope retainer
- boom hinge pin
- boom head section
- auxiliary boom head
- structure

10.0 Power Plant (Upper):
- exhaust system/guards/insulators
- belts/hoses
- guards/covers/rotating and reciprocating parts

11.0 Manual Section:
- alignment
- locking device
- structure

12.0 Lattice Boom Extension:
- boom, extension alignment
- cords
- lattices
- end connections
- storage device
- sheave(s)
- wire rope retainer
- structure

13.0 Jib:
- positive stops
- sheave(s)
- wire rope retainer(s)
- structure

PREVENTIVE MAINTENANCE CHECKLIST (PAGE 3 OF 3)	
14.0 Main Load Block & Hook:	**15.0 Overhaul Ball & Hook:**
• capacity marking • weight marking • sheaves • safety latches • 10° hook twist • 15% hook throat opening • 10% hook wear • swivel • bearing • wedge socket/end fitting • reeving • NDT results	• capacity marking • weight marking • safety latches • 10° hook twist • 15% hook throat opening • 10% hook wear • swivel • bearing • wedge socket/end fitting • NDT results

Note: Mark as checked each item inspected.

Defects Noted:

Item No.	Part Description	Remark

Recommendation

_____

_____

_____

_____

Supervise By: _____   Certified By: _____

Part V

Process Plants

Construction Process Plants are an integral part of a large construction project that needs aggregates, cement and asphalt materials. The economic advantage is much favorable for large projects such as roads, bridges and even buildings and dams to have its own supply plant. Due to its complex set up, its operations and maintenance should be handled by a separate crew of engineers and operators. One must have high standard of operating and stocking facilities at the project site. Important issue is to keep dust and spillage under control to keep the output of superior quality.

Construction Process Plants are feasible for acquisition and development as an independent project unit and are finance sustainable.

Project potential savings can be summarized as follows:

1. Reduce operating costs and transport costs.
2. Increase and improve in product quality and reduction of unwanted by-products.
3. Direct control of required production volume or quantity with on-site measurements of materials when you need them with no delays or waiting or additional time needed for deliveries.
4. Reduction of any wastages with control of production output exact project quantity at all time.
5. Increase up-time and eliminate additional ordering stages and time in ordering and deliveries of process materials.

DOI: 10.1201/9781003360667-18

14 Three Types of Construction Process Plants

14.1 CRUSHING PLANT

A general name of an equipment used for crushing stone is called Crusher. With these machines, the large stones are broken into smaller sizes and reduced to raw materials mostly for concrete mix and also for the bearing capacity of roads, railways, airfields, some bridges and infrastructure.

Crushers are widely used as a primary stage to produce the particulate product finer than about 50–100 mm.

They are classified as jaw, gyratory and cone crushers based on compression, cutter mill based on shear and hammer crusher based on impact.

14.1.1 CRUSHERS UTILIZATION BY PROJECT AREA

1. Stationary crusher is preferred in mines, quarries area where rocks are extracted first and have the largest size. The crushed stones are crushed to size according to construction requirements in various sizes.
2. Mobile crushers are generally preferred in construction sites in areas where available stones are of smaller sizes so that specific sizes of stones can be provided for immediate usage. When the construction is finished, it is feasible to move the equipment to the next project. So, mobile crushers are much faster and functional.

14.1.2 CRUSHER BY EQUIPMENT TYPE

1. Gyratory crushers are typically only used if the targeted amount of production is more than 1 million tons per year. These machines are made-to-order. A gyratory crusher includes a solid cone set on a revolving shaft and placed within a hollow body, which has conical or vertical sloping sides. Material is crushed when the crushing surfaces approach each other and the crushed products fall through the discharging opening.
2. Jaw crushers are workhorse crushers and the most commonly used. A jaw crusher consists essentially of two crushing plates, inclined to each other forming a horizontal opening by their lower borders. Material is crushed between a fixed and a movable plate by reciprocating pressure

DOI: 10.1201/9781003360667-19

until the crushed product becomes small enough to pass through the gap between the crushing plates. A jaw crusher can be used for maximal feed size 95 mm, final fineness (depends on gap setting) 0.3–15 mm and maximal continuous throughput 250 kg/h.

3. Cone crushers are popular rock crushing machines in aggregates production, mining operations and recycling applications. They are normally used in secondary, tertiary and quaternary crushing stages. However, in cases where the grain size of the processed material is small enough by nature and the traditional primary crushing stage is not needed, also cone crushers can carry out the first stage of the crushing process.

4. Impact crushers are best for less abrasive material, like limestone. These crushers invoke higher wear costs over time but generally provide greater reduction and possibly better shape.

The two main components to consider when identifying purpose are feed size and type of material to be crushed. There are a variety of other factors to review as well – the long-term strategy for the site, discharge gradation, moisture content of the material and the targeted amount of production, just to name a few.

14.2   CEMENT PLANT

Construction projects use the batching plant or the concrete batching plant for their cement requirements. Various equipment are combined together to mix concrete ingredients to form ready-to-use cement for the project. Some of these inputs include water, air, admixtures, sand, aggregate (rocks, gravel, etc.), fly ash, silica fume, slag and cement.

The major parts of a concrete plants are: mixers (either *tilt drum* or *horizontal*, or in some cases both), cement batchers, aggregate batchers, conveyors, radial stackers, aggregate bins, cement bins, heaters, chillers, cement silos, batch plant controls and dust collectors.

The most important equipment is the *mixer*, and there are many types of mixers, such as tilt drum, pan, planetary, single shaft and twin shaft. The twin shaft mixer can ensure an even mixture of concrete through the use of high horsepower motors, while the tilt mixer offers a comparatively large batch of concrete mix (see Table 14.1).

14.2.1   Types of Concrete Plants

For construction projects, concrete plants can be divided into dry mix plant and wet mixing plants, depending on whether a central mixer is used. They can be also divided into stationary concrete plants and mobile concrete plants, depending on whether they can be moved.

TABLE 14.1

A complete typical set of specification range of a concrete plant

Item	Capacity Ranges
Productivity	25–240 m³/h
Mixer discharging capacity	0.5–4 m³
Feeding Mode	Lift hopper and belt conveyor
Kinds of aggregate	2/3/4–3/4/5
Discharging height	3.8–4.3 m
Weight	10 t–98 t
General installed power	≈65 kw–≈340 kw

14.2.1.1 Dry Mix Concrete Plant

A *dry mix concrete plant,* also known as a *transit mix plant,* weighs sand, gravel and cement in weigh batchers via digital or manual scales. All the ingredients are then discharged into a chute, which discharges into a truck. Meanwhile, water is either being weighed or volumetrically metered and discharged through the same charging chute into the mixer truck.

These ingredients are then mixed for a minimum of 70–100 revolutions during transportation to the project jobsite. Dry mix plants typically see more break strength standard deviation and variation from load to load because of inconsistencies in mix times, truck blade and drum conditions, traffic conditions, etc.

14.2.1.2 Wet Mix Concrete Plant

A *wet mix concrete plant* combines some or all of the above ingredients (including water) into a central concrete mixer – that is, the concrete is mixed at a single point and then simply agitated on the way to the jobsite to prevent setting (using agitators or ready mix trucks) or hauled to the jobsite in an open-bodied dump truck.

Wet mix plants contain a central mixer, which can offer a more consistent mixture in a shorter time (generally 5 minutes or less). With a central mix plant, all loads are accurately and consistently measured for every batch; therefore, there is the same mixed output each time and there is an initial quality control point when discharging from the central mixer.

14.2.1.3 Stationary Concrete Batching Plant

Concrete batching plant is a fully automatic concrete mixing equipment consisting of the five major systems such as mixer, material weighing system, material conveying system, material storage system and control system and other ancillary facilities.

14.2.1.4 Mobile Ready-Mix Batch Plants (Truck Mounted)

The basic type of concrete mix plant is by a concrete truck mixer. The ingredients for the concrete mixture (without water) are combined using a ready-mix concrete batch plant, and the dry mixture is disposed of into a concrete transport truck where water is added. This gets the concrete ready for production. This mixing process occurs when the truck is on its way to the job site.

There are basically three different options that you have if you are going to be using ready-mixed concrete:

1. The concrete mixer can be turned at slow speed while inside the drum during transit, at which point the speed for the mixing drum can be increased for 5 minutes to prepare the mixture.
2. The concrete mixture can be mixed at the mixing yard and only slowly agitated during transportation to the job site.
3. The concrete mixture can be turned at medium speed while inside the drum during transit so that it can be completely mixed by the time it reaches its destination.

14.2.2　Cement Plant Application

Typical concrete plants are used for ready mix, civil infrastructure and precast applications.

14.2.2.1　For Ready Mix

A ready mix concrete plant is generally located inside the city, transporting ready-mixed concrete for projects through concrete truck mixers. Ready mix concrete plants have higher requirements for durability, reliability, safety and environmental protection of the concrete plant's system than other types of plant.

14.2.2.2　For Precast Applications

Precast concrete, also named PC component, is a concrete product that is processed in a standardized process in the factory. Compared with cast-in-place concrete, precast concrete can be produced, poured and cured in batches. A precast concrete batching plant has a safer construction environment, lower cost and high-quality products compared with concrete poured on site; the construction speed can be guaranteed. In addition, it is widely used in transportation, construction, water conservancy and other fields.

Precast and prestress concrete producers supply critical elements used in worldwide infrastructure, including buildings, bridges, parking decks, road surfaces and retaining walls.

14.2.3　Plant Dust And Water Pollution Control

Municipalities, especially in urban or residential areas, have been concerned by pollution from concrete batching plants. The absence of suitable dust collection and filter systems in cement silos or at the truck loading point is the major source of particulate matter emission in the air.

The loading point is a large emission point for dust pollution, so many project operations use central dust collectors to contain this dust. Notably, many transit mix (dry loading) plants create significantly more dust pollution than central mix plants due to the nature of the batching process. A final source of concern for many municipalities is the presence of extensive water runoff and reuse for water spilled on a producer's sites. Project safety unit must give some concern for environmental considerations.

14.3 ASPHALT PLANT

The plant blends aggregates and bitumen to produce the asphalt mix. This mix is then enhanced with additives and mineral fillers, and the mix is then applied for the construction of pavements of highways, parking lots, airport expressways, roads and dams.

Asphalt mixing plants are mainly composed of a batching system, drying system, combustion system, hot material lifting, vibrating screen, hot material storage bin, weighing and mixing system, asphalt supply system, power supply system, dust removal system, finished product silo and control system.

14.3.1 Types Of Asphalt Batch Plants

14.3.1.1 Based on Production Capacity

The small and medium asphalt plants are typically used for smaller construction engineering. These include asphalt mixing plants of capacity from 20 TPH to 100 TPH. They are used for road construction, parking lots, roads and so forth. The cost of transportation will be less than the cost of mobilization.

The large and mobile asphalt mixing plants are suitable for construction projects which require a substantial amount of hot mix asphalt. These are commonly used for construction purposes in metropolitan areas. They have a capacity of more than 100 TPH. To be noted is the fact that a mobile asphalt plant is more expensive than a stationary one for the primary reason that the plant and all accessories are made into as assemble and disassemble mobile mode.

14.3.1.2 Based on the Mobility

The mobile asphalt mix plants are becoming increasingly popular due to their movable aspect. The equipment set is compacted and is mounted on a chassis which moves around and aids in the transportation of the mixture. They are available in capacities ranging from 20 TPH to 120 TPH.

14.3.1.3 Based on the Technical Process

The continuous asphalt drum mix plants are capable of producing asphalt mixture steadily without any interruptions. They can integrate the drying and asphalt mixing process together at relatively lower cost. It is favored in large construction sites.

The asphalt batch mix plants are widely used for large construction projects. It can produce the highest quality of asphalt mixture. It is best suited for those projects which require the specifications of the mixture to be changed during the process.

14.3.1.4   Asphalt Recycling Plants

These plants can recycle reclaimed asphalt pavement using hot recycling technology. Post recycling, they produce a fresh batch of asphalt mix that can be used for small construction purposes.

Note: Pollution control: Asphalt plants need to adhere to changing pollution control standards, else could result in close down of the plant, causing enormous economic losses. This is an area that needs to be kept in mind in terms of customization.

14.3.2   ASPHALT PLANT CLASSIFICATION BASED ON EQUIPMENT USED

The basic purpose of all of these types is to produce hot mix asphalt. However, there are key differences between these plants in terms of the way in which they achieve the desired specifications and in overall working operations.

14.3.2.1   Batch Mix Plant

There are several aspects involved in an asphalt concrete batch mix plant. One feature of such plants is the use of cold aggregate feeder bins to store and feed the aggregates in different components according to their sizes.

Conveyor is used to shift aggregates from one place to another. Ultimately, all of the material is transferred to the drying drum. However, the aggregates also have to go through the vibrating screen to ensure the proper removal of oversized materials.

The drying drum consists of a burner unit to remove moisture and heat up the aggregates to warrant an optimum mixing temperature. An elevator is used to carry the aggregates to the top of the tower. The tower has three main units: a vibrating screen, hot bins and the mixing unit. Once the aggregates are separated by the vibrating screen according to their size, they are temporarily stored into various compartments called hot bins.

From the hot bins storage, the aggregates are released into the mixing unit. When the aggregates are weighed and released, bitumen and other essential materials are often released into the mixing unit.

In most industrial sectors, installing air pollution control devices is essential to ensure the sustainability and eco-friendliness of the asphalt plants. Typically, bag filter units are used to trap the dust particles. The dust is often reused in the aggregate elevator.

14.3.2.2   Drum Mix Plant

Drum mix asphalt plants have a lot of similarities to batch mix plants. Cold bins are used in drum mix plants. Moreover, the process is identical to the batch mix plant until the aggregates enter the drum after going through the vibrating screen to separate them on the basis of their sizes.

The drum has two main functions: drying and mixing. The first part of the drum is used to heat the aggregates. Secondly, aggregates are mixed with bitumen and other filter material. It is important to note that the drum mix asphalt plant is a continuous mixing plant. Therefore, small-size containers of a suitable material are used to hold the hot mix asphalt.

Since bitumen is mixed at a later stage of the production, it is first stored in separate tanks and then inserted into the second part of the drum. It is important to maintain optimum air quality to avoid pollution. For this purpose, pollution control devices like wet scrubbers or bag filters are typically used in drum mix asphalt plants.

It is evident that both types of plants have some common components and working procedures. For instance, feed bins are essential in both batch and continuous plants. Similarly, a vibrating screen is important in every type of asphalt plant. Other parts of the plants like bucket elevators, mixing units like drums, weighing hoppers, storage tanks, bag filters and control cabin are also important in both batch mix plant and drum mix plant.

14.3.2.3 The Bottom Line

The purpose of differentiating between these two major types of asphalt plants is to show that both types of plants produce good-quality hot mix asphalts, even if they do use different operating systems.

The type of asphalt plant that a company wants to set up is highly dependent on the construction project specifications, economic evaluation and the overall rules and regulations of the industrial area.

15 Process Plants Maintenance

Maintenance of any asphalt plant has to be ongoing and regular work. Neglected or sub-par maintenance on any plant components can lead to numerous break-downs and repairs.

Proper plant maintenance can be instrumental to the performance of your plant's next construction period. It has to be inspected as often to spot warning signs for technical problems in plant components during the operation period. Faulty parts on a plant can lead to more critical repair if not addressed when first seen. These then are more difficult to repair, even in the off-season.

Without proper maintenance, operating costs will be higher and incrementally lowers production output. Inefficiencies start to mount to the point that the plant is not keeping up with the project schedule. Breakdowns start to become more frequent, which further hinders an asphalt supply to keep up with the project schedule. Proper maintenance can avert such a situation.

15.1 MAINTENANCE TIPS FOR CRUSHERS

Crushers are machines used to reduce the size of rocks, stones and ore. They are used in aggregates production, construction material recycling and as part of the cement mix. Here we have maintenance tips for the different gyratory crushers, jaw crushers, cone crushers and mill crushers to give the maintenance care for these crushers.

15.1.1 JAW CRUSHERS

1. Check oil regularly, making sure it is free of dirt and contaminants.
2. Check jaw dies every day and make sure bolts are fastened securely.
3. Maintain crusher backing to ensure proper vibration and impact protection.
4. Keep breathers clean.
5. Make sure the toggle area is clean, especially prior to any adjustments.
6. Wash out seats and plates daily.

15.1.2 GYRATORY CRUSHERS

1. Check spider bushing clearance monthly, every 500 hours, or with every mantle change, whichever comes first.

2. Monitor and maintain bushing lubrication levels every day (about 1.4″ above the bushing flange). Make sure "extreme pressure" additive is added.
3. Check balance cylinder(s) every 30 days of operation.
4. Keep oil and lubricating points clean. Check daily, every 8 hours.
5. Follow crusher backing maintenance schedule carefully to avoid undue damage to equipment.
6. Every 40 hours, check for leaks and inspect Y strainers.
7. Every month, inspect safety devices and perform an oil analysis.
8. At least once per year (or every 2,000 hours) change all oil.
9. Know and use the proper mantle for your crusher. Failing to do so can result in serious damage.

15.1.3   Cone Crushers

1. Keep cone crushers choke-fed during operation. This will prolong the manganese service life and maintain the quality shape of the product.
2. Maintain oil cleanliness. Check daily.
3. Lubricate bearings every 500 hours and change the oil every 6,000 hours.
4. Make sure crusher backing is applied regularly to protect parts and components and fill in gaps.

15.1.4   Grinding Mill Crushers

1. Keep your equipment clean, preferable to clean after each use. From heads to cables and hoses to the remote-control unit, keeping your grinder clean will give you optimal usage for the entire life of the machine.
2. Check for uneven wear on tooling, cracks or other damage. This may be an indication of a damaged tool holder.
3. Use Original Equipment Manufacturer (OEM) parts only.
4. Pay attention to strange sounds, scratch patterns, power loss or anything else out of the ordinary.
5. Perform a deep clean on your mill weekly, breaking down the tips, checking for wear and erratic oscillation, belt slippage and proper battery maintenance if applicable.
6. Every 100 hours execute a full breakdown and inspection, being sure to clean the mill and make sure everything is in proper working order.
7. A yearly tune-up from a professional service provider will give you the edge when it comes to keeping your machinery up and running well.

Given the high-impact stresses experienced by crushing equipment, even slight damage can subsequently cause damage from one part of the crusher to another. Keeping up with your crusher maintenance schedule is crucial to keeping your equipment running and avoiding unexpected downtime.

15.2   MAINTENANCE TIPS FOR CONCRETE MIXING PLANT

Concrete mixing plants are more demanding to maintain just like asphalt plants due to the high moisture content nature of their products. Given are some of the most important steps for equipment quality outputs.

The following tips are general recommendations suitable for most industrial concrete mixers. Mixer operators need to read the operation and maintenance manual supplied with their mixer before the operation begins. The concrete mixer service and maintenance tips below may differ from your specific product recommendations. Always follow your manufacturer's recommendations for specific design differences.

15.2.1   DAILY MAINTENANCE

1. The check mixer is empty at the end of each operation and before the next shift starts.
2. Make sure there is no waste concrete on the floor or walls of the mixer.
3. Operate high-pressure washing each time there is an extended pause in production.
4. Check to see that the discharge chute is also clean.
5. Check cement wet hopper discharge valve is free of buildup through the inspection panel at the base of the weight hopper.
6. When cleaning any buildup on mixer arms, mixing blades, scraper blades and wall and floor tiles, do not use metal hammers because these items are cast iron and therefore brittle. Use a rubber head hammer.
7. Make sure the door seal is free of track buildup.

15.2.2   WEEKLY MAINTENANCE

1. Check the clearance between the mixing blade and the floor of the mixer. Adjust the clearance to approximately 3 mm, if necessary, by adjusting the height of the mixer blade.
2. Check the clearance between the wall scraper blade and the wall tiles. Adjust the wall tiles on the perimeter to be clean.
3. After adjustment, rotate the mixer by hand to check the clearance on all floor and wall tiles before starting the machine. Always check bolt tightness after making any adjustments.
4. Grease all nipples on top of the fixed gearbox. Do not apply too much oil.
5. Check the hydraulic oil level in the power pack tank using a vision glass. This should not require topping up unless there is a leak in the system.
6. Check drive belt tension and high-pressure washer condition and adjust if necessary.

7. Check for cracks or damage to wall and floor tiles. Casting blowholes can be ignored. When the tile thickness reaches approximately 5 mm, replace the entire set of wall or floor tiles as needed.
8. Use a wire brush to remove high-pressure washer spray heads and cement/concrete buildup that may be blocking the nozzle.

15.2.3 MONTHLY MAINTENANCE

1. Check the condition of the discharge door seal and replace if leaking.
2. Check the oil level in the gearbox. Use the dipstick on the fixed gearbox and the overflow plug on the rotating gearbox. Do not overfill any box. The oil level in the fixed box should be up to the dipstick base.
3. While the oil may seem emulsifying, it is not water contaminated and is a normal appearance for this oil. Oil in the gearbox should not be used and a drop in level indicates a leak and should be checked at once.
4. Check the oil level on the high-pressure water pump. Top up as needed to see gauge marker.

15.2.4 MAINTENANCE TIPS FOR CONCRETE TRUCK MIXER

1. Clean the mixer daily. Make sure drum and mixing paddle are free of any bits of cement built ups, clean it with a water hose to wash away cement bits before there is any concrete buildup.
2. In case there is any concrete buildup, you have to use the pressure washer to get rid of it. If it doesn't work, you have to get rid of the buildup by chipping.
3. You need to keep the motor clean; you can use an air compressor to blow out cement and any other dust particle. By keeping the motor clean, you will make sure it performs the best. In addition, you are also keeping yourself safe from costly repairs. Remember, the loose particles can be troublesome for your motor as they can lead to expensive repairs.
4. You have to grease the parts and pulley of your mixer. It will prevent damage from friction. This extends the lifespan of your device and ensures it performs the best

15.3 MAINTENANCE TIPS FOR ASPHALT PLANT

Each type of asphalt plant has its own unique design and therefore different operation and maintenance procedures. Here we have important maintenance steps by main components.

15.3.1 COLD FEED AND CONVEYING SYSTEM

1. Proper belt tracking is crucial on weighbridge conveyors. This should be monitored daily during production. A belt that is not running straight can be quickly damaged and require correction or replacement.

2. Check conveyor components (i.e., troughing rollers, return rollers and belting) for excessive wear, damage and material buildup. Replace worn and broken components. These can lead to piles of aggregate which require considerable time and effort to remove by hand.
3. Examine the conveyor's bearings and drive system. For chain drives, examine each sprocket for "fish eyeing", marking any abnormalities you find.
4. Check for cracked or glazed belts, excessively worn sheaves add a loose condition which may require adjustment.

15.3.2 DRUM

1. Important to focus maintenance on asphalt plant drum.
   a. Check drum shell thickness for wear.
   b. Clean, check and tune the burner.
   c. Change oil in drive reducers. Check the bag house to ensure bags are in good shape. Also ensure cleaning and fines-return systems are working properly.
   d. Check trunnions for excessive wear or uneven wear, and check the bearings for excessive play.
   e. For internals, the areas which should be monitored include the flighting and the shell (mapping its thickness, particularly for parts with high wear incidents).
2. Repair and replacement of components will minimize fuel consumption and maximize the life of the dryer shell. Specifically, the combustion flights can make a substantial impact on both the fuel combustion efficiency and emissions.
3. For external of the dryer/drum mixer, the tires and trunnions should be monitored to make sure that everything is lined up and parallel and that there is minimal wear and proper alignment. Repaired or replaced if it has wear beyond the hardening point or have severe metal flaking.
4. The tires and trunnions should be ground or polished by properly trained field technicians. Also on the external, some drum designs require massive pugmill arms and mixers on the outside of the shell. These must be maintained on a systematic basis, including digging out by jackhammering and cleaning.
5. All mixing areas, whether it is a pug mill on a batch plant or by other mixers, need to be inspected, wear-checked, repaired and restored to proper condition.

15.3.3 SCALPING SCREEN SYSTEMS

In general, check and inspect for impaired components, worn sheaves or belts and broken motor mounts.

The following steps are recommended.

1. Scalping screens are important in removing (rejecting) oversized aggregate for both batch and continuous mix asphalt plants. What goes through the plant determines the final product. Start with an external inspection.
   a. Check the size of the feed material. If it's a useable size that normally goes into a hot bin, the accumulation could point to leaks in the oversize discharge, exacerbated by screen flooding and carry-over.
   b. Check the following parts: the skirting seals under the units, the top covers for wear, the screens are moving freely, under the drive end for built up of aggregates.
   c. Inspect and check for any missing parts, such as lid hold-downs, and for signs of dust leaks.
2. Remove all the lids and side covers. Use a hose and clean the accumulated fines off the screen frames and springs to make it easier to find damaged and cracked components. Caution: Remember to open all the gates below and provide a way for the water to escape from under the plant to a waste water basin.
   After cleaning inspect for broken springs and cracks in the framework.
3. Closely check and inspect each screen cloth. Check for excessive wear and broken wire. Check the screen trays. All must be tight. Check screen cloth security.
4. Stock up a complete change of screens. When stored in a custom-built rack on the screen deck, the installation will be with minimum downtime. Keep the appropriate bolts handy.
5. Examine the screen drive and eccentric. Follow the OEM's recommendations as to periodic maintenance which is according to the materials used.

15.3.4   BATCHER/SILO SYSTEMS

The batcher and silo systems are two of the most important pieces of equipment on an asphalt plant when it comes to battling mix segregation and safely storing your mix (Figure 15.1).

1. The silo cone should be thickness-mapped on an annual basis, at a minimum, and perhaps more depending on the amount of tonnage. Check thickness gauge readings on the length and circumference of the cone section and up the silo sidewalls, depending on how the silo is operated, and measure the amount it is filled and pulled down below the cone level on a daily basis. This measurement and mapping should be done for each silo on a routine basis to closely monitor how much wear has occurred (Figure 15.2).
2. When a batcher unit has problems, its gates are chained open to ensure the job moves forward. The asphalt mix suffers, but the value of a

FIGURE 15.1 Process plant silo.

FIGURE 15.2 Concrete truck mixer.

FIGURE 15.3 Dump truck loading aggregate materials.

shutdown for repairs to the more batch is deemed excessive and as long as the mix is accepted. Skip but go back after full usage (Figures 15.3 and 15.4).

Inspect the batcher's sides and cone for thin spots. Mark for any plating that may be required. Look closely at the gates and their pivots. Are they loose? Do they need to be greased? Mark for repairs any abnormalities you find.

FIGURE 15.4 Asphalt paver operation.

3. Examine the air actuation system. Failures here are the primary cause of more batch problems. Check the air cylinder(s) for excessive wear. Replace worn and substandard hose.

4. Use the same inspection protocol for the storage silos as for the batchers. Check the structural thickness, especially in the cone area. Mark any thin areas for repair or replacement. Most manufacturers offer onsite density inspections to determine wall thickness.

5. The gate and all its actuating controls should be examined. Check the pivot pins by prying the gates around with a bar. Schedule repairs for any found lacking. Check the operation of the silo's high- and low-level warning alarms.

15.3.5 Dryer/Mixer System

Dryer flighting and mixer tip condition are imperative for adequate aggregate moisture removal and thorough blending.

1. Conditioning, showering and combustion flights need to be inspected for wear or damage. Replace or repair as needed. Inspect each and every flight and mixing tip for the following problems:

   1. Excessive wear
   2. Heat curling
   3. Impact bending
   4. Cracking or absence

2. Mark for a replacement for any flight that exhibits these problems. Repair and weld any that are cracked can be welded, and those that are worn or missing must be replaced. Mixing tip wear needs to be evaluated periodically or when the blending quality of the asphalt mix appears inadequate.

3. Drum buildup removal should be performed in conjunction with rotating mixing tips (paddles).

15.3.6 Combustion Systems

The main burner and hot oil heaters make up your combustion system. Burner maintenance/cleaning and periodic tuning are important to utilize every British Thermal Unit (BTU) fuel heat properly.

1. Test, evaluate and tune your burner every 100 hours to optimize performance. Having a properly running burner must conform to local government regulations.

2. Examine all fuel and propane lines. Schedule for replacement of any that gives irregular operations. Check the propane bottle for signs of leakage or damage. Check the fuel pump and drive mechanism. Mark any frayed belts, worn couplers or leaking seals. Change the fuel filters regularly or every 100 hours, for replacement.

3. Check the piping and valves from the supply tank. Also, take a minute and inspect the tank's filling apparatus. Note any leaks or unsafe conditions.

The maintenance of the burners will depend on the fuel type used. Whether the burner is using fuel oil or gas, regularly inspect and clean the fuel delivery system. Also check, inspect and clean regularly the pilot assembly and burner nozzles. If equipped with electric heat for your hot oil, check the operation regularly.

15.4   PLANT GENERAL SAFETY MEASURES

1. Safety precautions must be a daily routine during operation periods.
2. Emergency shutdown switches, safety disconnects on conveyor and turn head motors should be included in plant set up.
3. Every person working in plant should wear safety accessories such as gloves, apron, helmet, goggle, ear cover and safety shoe.
4. Handrails must be provided for stairs, ladders and top floor levels.
5. Electrical wiring should be inspected regularly.
6. Entry access into the manholes of cement or aggregate compartments must be limited.

Part VI

Equipment Safety and Security

16 Equipment Safety

16.1 SAFETY FUNCTION AND ORGANIZATION

Safety is paramount for operation like heavy equipment operation. Management must be aware of the intrinsic and extrinsic factors emanating from ownership of heavy equipment. Intrinsic are those inherent and coming from the equipment design system or of an assembly or lacking thereof in the equipment or within effects that originate from any part of the equipment.

Extrinsic factors are those ensuing from the usage of the equipment such as manpower, maintenance and environment conditions. Examples of intrinsic effect in nature are like mechanical failure, disregard or lack of equipment safety design, and examples of extrinsic effect are environmental effect and human or operator error.

The primary goal of safety is to manage risk and eliminate or reduce accidents emanating from one or both factors to attain zero or minimum accident. Equipment manufacturer nowadays is very concern of their equipment safety features and make it a point of their sale. From engine safety operation and design to hydraulic safety gadgets and to simple but indispensable safety of view mirrors, back horn, night light, tank caps and flame arrestor, boom angle and length indicators for cranes, even air conditioning for operator's protection and physical welfare.

Some extra safety features are mandatory for some countries such as air condition in desert projects against heat and sand storm but not in tropical project where a good canopy will suffice.

In purchasing equipment, management must ask from manufacturer or distributor what safety features does equipment have and get detailed safety features of engine, transmission, clutch and brake, hydraulics and electrical set up as part of selection process. This will address the intrinsic element.

Another important step to safety is to contact the safety organization in the country of operation and have as many as possible manuals listing the codes and guidelines of the country. For heavy equipment, some are part of the construction safety association or can be alone as equipment association. Organizations working toward improving construction safety include:

- Construction Safety Council
- ConstructionSafety.org
- CPWR – The Center for Construction Research and Training

For the extrinsic elements, safety must start from hiring process to having an organization for safety in the job site or project. First safety must be addressed in manpower selection for the right trained and experienced personnel for each type

DOI: 10.1201/9781003360667-22

of equipment with emphasis on heavy lifting equipment personnel. Qualification standards must be established and made part of the testing of operators prior to hiring. Second will be the safety organization at the job site.

A safety engineer or supervisor must be assigned in the project. His main job is to increase attention to safety and positive action by all in the job site directly or indirectly associated with equipment operation.

Due to the complexity of a construction site, safety must always guard against hidden dangers by taking proper precautions. For example, heavy equipment, such as cranes and forklifts, requires experienced and competent operators. Safety must identify every hazardous materials on-site and clearly mark them with the appropriate warning signs. Train workers who use hazardous materials to deal with any emergency situations.

Sign must be set up near any large openings in floors to prevent falls, and set up guardrails in high areas that do not have walls. To ensure capturing all possible hazards in the project site, have the guide below as initial input to organize the safety set up.

Typical safety concerns of the safety organization are (Figure 16.1):

1. Equipment and machine compliance to safety operation particularly cranes and lifting equipment.
2. Personal protective gear and implementation.
3. Employee involvement to safety practices starts in knowing their qualification.
4. Regular safety inspection to ensure safety compliance.
5. Identify workplace hazards or anticipate danger to equipment.
6. Chemical and health hazardous materials in the job site.
7. Accident prevention and safety promotion in the project.
8. Accident and emergency response availability.
9. Short but concise safety meeting monthly for all sectors in the job site.

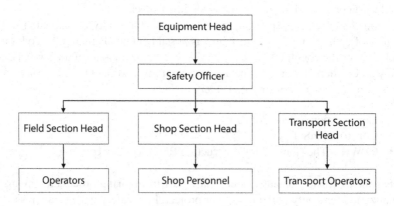

FIGURE 16.1 Typical safety organization set up at the project site.

Equipment accident preventive plans:

1. Equipment accidents are preventable, given and following the good precautionary procedures in the project sites and shops during operation times. Equipment manager and engineers must provide the safety guidelines for each equipment in the construction sites.
2. Personnel training and orientation prior to hiring and assignment to operate an equipment must be done. These are procedures on handling controls such as use of back-up alarms, negotiating stiff slopes, proper mounting and dismounting of personnel and loads and mandatory use of seatbelts. Manufacturer operation manuals should be used in the training and must be available at all times.
3. Daily inspection of equipment prior to operating and performance of assignment must be done. Inspection sheets must be given to operators to ensure mastery of his equipment of the operator.
4. Parking of equipment must be provided far from site offices and construction sites. This must be on level ground and well drained.
5. Follow the procedure for equipment moth balling in case equipment shall be parked for long period of time.

The safety organization shall be the direct link of the safety officer to all section heads, who in turn shall relate the safety orders of the officer to his subordinates. While this is not a line function, it is an administrative function that gives the safety officer authority to instruct section heads on safety concerns in the project.

16.1.1 SAFETY REPORT

To keep management abreast on safety, it should receive regular safety report primary from the monthly meeting of supervisors chaired by the safety officer of the project. It shall contain discussion imminent safety problems and items in the project. The guide line can use the nine concern items listed above as checklist.

The report shall be prepared by the safety engineer or supervisor and submitted to the project manager. More details can be developed and expanded as the project goes on, particularly on items inherent to the project type, location and condition.

16.1.2 TYPICAL ACCIDENTS OF HEAVY EQUIPMENT

1. Caught between vehicles backing up, especially for large on-site vehicles with huge blind spots.
2. Pinned by the heavy equipment body, loads or implements.
3. Hit by load or cargo during lifting and shifting of position.
4. Caught and dragged along by moving or rolling equipment.
5. Crushed under collapsing structures.

6. Stumbling during mounting or dismounting implements, tires and loads of the equipment.
7. Equipment collation in blind corners and overcast environments such as heavy rain, sandstorms or simply dust from the construction site.

16.1.3 Typical Crawler Tractor Safety Operating Procedures

Crawler tractor or bulldozer is one of the most common heavy equipment used in construction projects. While it is simple to operate, safety precautions like any other on- and off-road equipment must be emphasized in the project site. Here are some very typical safety operation procedures for crawler tractors or bulldozers.

1. Before operating a crawler tractor, check the work area for obstructions and hazards including ditches, slopes, hills, excavations, streams and underground and aboveground utility lines.
2. Bulldozer operators must wear flagging garments (i.e., orange vests) and, at least, a hard hat, steel-toed boots, long pants and glasses against dust or sandstorm if needed.
3. Never carry passengers when operating a crawler tractor and stunt driving is prohibited.
4. Do not operate a crawler tractor with a leaking fuel system or leaking brake system.
5. Crawler tractor must be equipped with rollover protection structures (ROPS).
6. Operators must always use seatbelts when operating a bulldozer.
7. Operators must mount and dismount using only the steps and handholds provided on the crawler tractor.
8. Crawler tractor blades and attachments must be kept close to the ground while moving.
9. Crawler tractor should be parked on level ground after use or during night parking, whenever possible.
10. The crawler tractor must be equipped with an automatic back-up alarm.
11. Never use a crawler tractor to demolish structures that are taller or heavier than the dozer itself, unless there is sufficient overhead protection.
12. Mechanics must follow proper lockout, blockout and tagout procedures when a crawler tractor bulldozer is in repair.
13. When parking, shift into neutral and set the parking brake to immobilize the dozer. Lower the blade and any other attachment to the ground and then shut off engine.
14. Operators must use proper towing procedures and equipment for crawler tractor attachments.

16.1.4 Mandatory Crane Equipment Safety Items

Other than the normal on-highway safety features and devices such as tire inflation, view mirrors, night light and back-up light mobile crane of both, truck and crawler cranes are equipped with extra safety devices akin to its operation.

These are boon angle indicators, boon length indicators, audible safety warning signal and horn, shock-absorbing boom stop and boom hoist shutoffs, automatic device to stop boom drum motion when maximum permissible boom angle is reached, spirit level at the outrigger controls for leveling the unit, machined surface on the revolving plate for precision leveling and plate installed on boom vicinity to indicate distance from center of rotation.

1. Check factors affecting capacity of the crane. The load capacity of the crane is based on ideal condition, and field actual working condition and application differ from project to project. First and foremost is to know the capacity from the chart and then check factors which may reduce these loads. Nevertheless always use the load chart, and do not rely on stability to determine lifting capability.
2. The boom load increases as the radius decreases. Likewise the boom loads drop as the radius is increased and then they rise again as they approach horizontal. This is shown in the component loading graph given for each crane type. Check this as basic reference.
3. Match the load carrying capacity of the crane with the boom length as given in the chart. There is a chart for the effect of boom length on capacity. Check this chart.
4. The capacity of a crane equipped with a jib is affected by length of main boom, boom angel, jib length, jib angle with respect to boom, weight of the rigging and both hooks and the mathematical computation of strength of jib, of the main boom and stability of the crane.
5. Mobile crane must never be used on soft or uneven grounds or on slopes where crane stability is compromised. Firm footing and uniform ground level should be provided in accordance with set-up procedure given with the equipment.
6. Safe working loads of machines equipped with outriggers are generally stated two ways. These are either "on tire or crawler" or "on outrigger". The outrigger rating is based on all outrigger beams fully extended, all tires or crawlers within the outrigger boundary being completely off ground. Unless these two conditions are met, the outrigger rating cannot be used and the safe working load drops to the on tires rating.
7. Rigging the load properly is as important as knowing the load. Know the safe working load of the tackle, and rigging equipment used must never exceed the limit. Check the machine requirements. Never use damaged sling and avoid sharp bends, pinching and crushing.
8. Go into details in ensuring of handling the load properly. From proper rigging to taglines control, ensure the boom peak is directly over the load, the load line is directly over the center of gravity of the load and the hoist line is vertical. No jerking or impact loading. No dragging of load. It is strongly advised that all mobile cranes be equipped with power load lowering devise for precise control in lowering of all loads.

9. Always provide the safety procedure and use a signal man to assist the operator.
10. When lifting with more than one crane, do a detailed cranes operation scheme beforehand. This is extremely complex procedure that need specific plan before execution to ensure safe and success of the operation.

16.1.5 JOB SITE CLINIC

Complimentary to the safety organization set-up is the establishment of a clinic in the project or job site. The clinic must have all the basic health and first aid requirements to attend any physical injury in the project. Project reaching 40 units shall have a full-time nurse during the operating shift in the project.

The project should also have the information on nearest other clinics and hospital in the area of operation. A service vehicle must be assigned as ambulance to transport any injured personnel to the clinic or nearest hospital for immediate medical attention in case of accident.

17 Security Organization

Equipment security is a major part of project security. It is the measures taken to safeguard the project office assets, technical documents and records, warehouse and supply inventory, equipment and machine assets and properties, and the construction on progress development. Depending on the job site, if it is in one location or in many simultaneous job locations, security must be planned and sustained until project completion.

Equipment is high-value assets and must be guarded against pilferage and malicious breakages from both external and internal sources. To guarantee security, have a security plan which is amalgamate to the project security plan. As a policy, the plan must be confidential and discussed with the project and equipment officials, who must ensure compliance in terms of manifesting the plan in its technical requirements such as manpower, paraphernalia and mapping out of project and job sites.

Where and how to park and store equipment, secure warehouse and office, limit people to secured areas and control incoming and outgoing of personnel and project guest in the job site? All this must be contained in a master security plan designed by a security management expert and the project officials including the equipment officials.

For the equipment operations, the areas to be considered are:

1. Electronic project files and communications facilities.
2. Warehouse and inventory supplies and access thereto.
3. Shop maintenance infrastructure and machineries and access thereto.
4. Equipment yard and access thereto.
5. Good lighting facilities particularly for and during increment weather condition.
6. Control of incoming and outgoing personnel and guest such as suppliers and visitors to the project premises. Not allowing loitering and unofficial stay in at night.
7. Check system of electrical installations, usage and after when project takes a break especially at night time.
8. Emergency drill and SOP for security to respond and execute in times of security alarm.

DOI: 10.1201/9781003360667-23

18 Company Equipment Library

The main function of a company equipment library is to organize and store all the equipment manuals and guidebooks which come with the brand new equipment purchase. The main materials in the library are the equipment maintenance and operation manuals and guidebooks. This must be given priority and main shelves for easy reference when demanded by users in the company (Figure 18.1).

Most libraries have materials arranged in a specified order according to a library classification system so that items may be located quickly and returned without misplacing a copy. The materials can be arranged by original equipment manufacturer, by operation and by maintenance and if available by parts book (Figure 18.2).

Some libraries have additional shelves where manuals on equipment performance, equipment references and catalogs of new equipment type, machines, accessories and tools in the market are stored. This must have separate shelves for good control.

Copies of the manuals and guidebooks on operations and maintenance must be available to the mechanics, electrician and technical manpower when needed for their reference and guide in doing maintenance and repair works. In instance when the book in needed at the project site, only the extra copy must be sent.

In case no extra copy is available, a photo copy of the portion and pages needed can be sent. This is to preserve one original copy at the library at all times. Request from the project site must be approved by the Equipment Superintendent and they must also be responsible for the return of the borrowed materials (Figure 18.3).

Parts and assembly books which are also issued upon purchase of new equipment will be registered at the library. The original copy will go to the Inventory Control section of the company. In case extra copies are available, the library must keep them for storage and safekeeping with the other same manufacturer group.

The library can also have the company computers for research and sourcing for materials related to equipment management and operations. It may order materials not available at the library, upon request from the equipment head of departments. Requisition of this kind must be channel to Purchasing section for approval. A permanent clerk must be made responsible to all library inventory and transactions (Figure 18.4).

DOI: 10.1201/9781003360667-24

FIGURE 18.1   Road safety barrels.

FIGURE 18.2   Road safety signs.

FIGURE 18.3   Safety road cones.

FIGURE 18.4   Safety construction vest.

Part VII

Equipment Rental Rate

19 Equipment Owning and Operating Cost

This chapter shall deal primarily on the equipment rental rate. Equipment owner is not only potential lessor but also lessee of equipment. In instance when there are surplus equipment, owner or management can take that opportunity to lease out their equipment after checking all particular factors such as the condition of the place of work as safe and the terms as feasible. On the other hand if in case project needs equipment for only a short application, then buying can prove uneconomical to own, so renting from outside source is the solution.

The rental rate formula most popular in use in the industry is the computation of the hourly owning and operating cost of equipment. The standard data used in the computation must be assessed in every assignment because of the many factors that can influence cost such as location of job site, intricacies of work on hand, local price of petroleum, oil, and lubricants (POL) interest charges, conduction permits and transportation or shipping to and back of equipment cost.

There are two main parts of the rental rate: (1) the fixed cost, which is also the owning cost, and (2) the variable cost which is the operating cost. This is the reason the rate is computed on a per hour basis (Figure 19.1).

Fixed cost is comprised of the depreciation value, operator or labor cost, tire or undercarriage cost, insurance, interest charges, taxes if any and miscellaneous and overhead cost. The equipment depreciation cost is the principal cost of owning cost and typically line depreciation method. Important here is the selection of depreciation period which is based on assets' write-off periods alone.

Owner must balance so as not to have an inflated value due to short depreciation period or to be undervalued due to an extended period of depreciation. The experience and recommendation of manufacturers usually provide the suggestion for the good full depreciation period for their equipment according to their type of machine (Figure 19.2).

Other fixed costs such as the operator or labor cost is the salary rate the personnel as given by accounting which should include benefits and training cost. Other fixed cost such as interest, taxes if any, insurance and miscellaneous can then be discerned from the establishment of the depreciation period.

PM-maintenance costs are cost allotment from the accounting department on total cost of maintenance over the life of the equipment. Tire or undercarriage cost can be computed according to the life of the tire or carriage in hours. Miscellaneous and overhead costs are factors to be taken from the accounting cost data as allotted for the management of the equipment fleet of the company.

The second part of the rate is the operating cost of the equipment. This is comprised largely of the fuel cost, lubricants, grease and filter cost. These are

DOI: 10.1201/9781003360667-26

FIGURE 19.1 Crawler excavator night operation.

FIGURE 19.2 Crawler shovel night operation.

TABLE 19.1
Typical Equipment Hourly Owning and Operating Cost Estimate

1.0 Fixed Cost (For Truck Tractor)

1.1 Depreciation Value	unit	Total	Remarks
a. Delivered Price with attachments:	$	240,000.	
b. Less: Resale Value	$	72,000.	Assumed approx. 30%
c. Net Value for Depreciation	$	168,000.	

1.2 Owning Cost

$$\frac{Net\ Value}{Total\ Hours} = \frac{168000}{12000}$$

	$\dfrac{\$}{Hr}$	14.00	Assumed total hours = 12,000

1.3 Interest, Insurance, Taxes

Interest = 15% \quad All assumed data.

Insurance = 3%

Taxes = 3%

Total = 21%

$$Net\ Value\ x\ Total \Big/ 2000 = \frac{\$168000x.21}{2000hr}$$

	$\dfrac{\$}{Hr}$	17.64	Assumed 2000 hr annual usage
Total Owning Cost (Fixed cost)	$\dfrac{\$}{Hr}$	31.64	

2.0 Variable Cost

2.1 Operating Cost

a. Operators Hourly Wage	$\dfrac{\$}{Hr}$	15.00	All data assumed
b. Fuel Cost	$\dfrac{\$}{Hr}$	8.00	All data assumed
Fuel Cost × Consumption			
$ 0.50/liter × 16 l/hr			
c. Lubrication & Filters	$\dfrac{\$}{Hr}$	0.52	All data assumed
Lube, Oil & Filter Cost × Consumption			
$ 0.04 × 13 (computed total units, reference use Cat performance guidebook)			
d. Tie Replacement Cost	$\dfrac{\$}{Hr}$	2.36	All data assumed

$$\frac{Tire\ Cost}{estimated\ life\ in\ hours} = \frac{\$5,200}{2,200\ hr.}$$

e. Repair Cost	$\dfrac{\$}{Hr}$	3.50	All data assumed

$$\frac{estimated\ annual\ repair\ cost}{total\ annual\ hours} = \frac{\$7000}{2000\ hr}$$

f. Attachment (if any)

Total Operating Cost (Variable Cost)	$\dfrac{\$}{Hr}$	29.38	All data assumed
Total Hourly Owning & Operating Cost	$\dfrac{\$}{Hr}$	61.02	

Note: Above data are all assumed and do not reflect actual cost to any specific equipment at any location. Cost of undercarriage shall replace cost of tires for crawler equipment.

consumables cost that is computed according to the number of hours the equipment is in operation.

Another extra cost of rental is the shipping and delivery cost of the hired equipment. This is chargeable directly to the lessee of the equipment according to the truck trailer rate and distance of the job site from central yard. If an equipment is rented with attachment, the depreciation cost in hours shall be charged to the user according the period of use of the attachment.

Most manufacturers like Caterpillar have in their performance manuals cost factors for each type of equipment as on highway or off highway and crawler equipment. It includes POL cost, accessories and tire or undercarriage cost among others.

The exercise of establishing equipment rental rate will give management or owners the confident to knowing the hourly cost of their equipment and plan their operation in detail (Table 19.1).

Bibliography

1. Equipment Operations and Technical Manual, by Ernesto A. Guzman, Philippine National Construction Corporation, May 1983
2. Crane Handbook, by D.E. Dickie, P. Eng., Construction Safety Association of Ontario, Canada, October 1975.
3. Caterpillar Handbook, Caterpillar Tractor co., CAT publication, edition 9.
4. Wikipedia Heavy Equipment http://en.wikipedia.org/wiki/Heavy_equipment.
5. Wikipedia Preventive Maintenance http://en.wikipedia.org/wiki/Preventive_Maintenance_Checks_and_Services
6. Wikipedia http://en.wikipedia.org/wiki/Maintenance
7. Wikipedia Lubrication http://en.wikipedia.org/wiki/Lubrication
8. Wikipedia Safety http://en.wikipedia.org/wiki/Safety_engineering
9. William Kent's Mechanical Engineers' Handbook – Twelfth Edition – Design and Power Volumes, Edited by J. Kenneth Salisbury.
10. Wikipedia High Bed Trailer|High Bed Trailer Manufacturers|High Bed Trailer Suppliers|High Bed Trailer Exporters.
11. Wikipedia Truck Tractor 2.jpg
12. Wikipedia Crane Load Charts, Brochures, and Specifications – CraneHunter.org
13. Wikipedia Equipment Details: 2005 FORD F650 FUEL & LUBE TRUCK.
14. Wikipedia Construction Equipment Parts & Attachments For Sale – New & Used Heavy Equipment Parts & Attachments at Rock & Dirt.
15. Wikipedia Link_Belt_RTC-8022_Specifications.png
16. Wikipedia On and Off Highway Tires.
17. Wikipedia: http://safety.ucanr.org/files/86423.pdf
18. http://images.search.yahoo.com/search/images;_ylt=A0SO8zEbd9JSKTkAT6dXNyoA;_ylu=X3oDMTB0cDAxNDRiBHNlYwNzYwRjb2xvA2dxMQR2dGlkA1ZJUDI5M18x?_adv_prop=image&fr=slv502-ytie&sz=all&va=heavy±equipment
19. http://images.search.yahoo.com/search/images;_ylt=A0SO8yAIe9JS9B4A86FXNyoA;_ylu=X3oDMTB0cDAxNDRiBHNlYwNzYwRjb2xvA2dxMQR2dGlkA1ZJUDI5M18x?_adv_prop=image&fr=slv502-ytie&sz=all&va=construction±crawler±crane
20. http://images.search.yahoo.com/search/images;_ylt=AwrTcX.WfNJStcsA6.2JzbkF?p=Road±construction±site&fr=slv502-ytie&ei=utf-8&n=60&x=wrt&y=Search
21. http://images.search.yahoo.com/search/images;_ylt=A0SO8w.NfNJStk4AK9JXNyoA;_ylu=X3oDMTB0cDAxNDRiBHNlYwNzYwRjb2xvA2dxMQR2dGlkA1ZJUDI5M18x?_adv_prop=image&fr=slv502-ytie&sz=all&va=construction±site
22. http://images.search.yahoo.com/search/images;_ylt=A0SO8zDsw9RSSxcA1YNXNyoA;_ylu=X3oDMTB0cDAxNDRiBHNlYwNzYwRjb2xvA2dxMQR2dGlkA1ZJUDI5M18x?_adv_prop=image&fr=slv502-ytie&sz=all&va=heavy±equipment
23. https://fabo.com.tr/en/what-is-a-crusher-plant-and-what-is-it-used-for/
24. https://www.zcrusher.com/jaw-crusher.html?gclid
25. https://www.zcrusher.com/jaw-crusher.html?gclid

26. https://www.mogroup.com/insights/blog/aggregates/choosing-the-right-primary-crusher-for-aggregate-production/
27. https://www.kaushikengineeringworks.com/guide-on-how-to-choose-the-right-asphalt-batch-plants
28. https://daswellmachinery.ph/seven-tips-for-choosing-right-concrete-batching-plant/
29. https://www.bigdreadymix.com/what-is-a-concrete-batch-plant/
30. https://www.kaushikengineeringworks.com/what-are-the-types-of-asphalt-plants/
31. https://www.thomasnet.com/insights/tips-for-crusher-maintenance
32. https://www.navyaindia.in/blog/Concrete-Mixing-Plant-Maintenance-Tips.html
33. https://www.forconstructionpros.com/asphalt/article/12098559/how-to-optimize-asphalt-plant-downtime
34. file:///C:/Users/Ernesto%20Guzman/OneDrive/Desktop/HE%20tower%20crane%20check%20list
35. https://essential.construction/academy/tutorials/a-guide-to-construction-tower-cranes/
36. https://en.wikipedia.org/wiki/Concrete_planthttps://fabo.com.tr/en/what-is-a-crusher-plant-and-what-is-it-used-for/
37. https://www.zcrusher.com/jaw-crusher.html?gclid
38. https://www.zcrusher.com/jaw-crusher.html?gclid
39. https://www.mogroup.com/insights/blog/aggregates/choosing-the-right-primary-crusher-for-aggregate-production/
40. https://www.kaushikengineeringworks.com/guide-on-how-to-choose-the-right-asphalt-batch-plants
41. https://daswellmachinery.ph/seven-tips-for-choosing-right-concrete-batching-plant/
42. https://www.bigdreadymix.com/what-is-a-concrete-batch-plant/
43. https://www.kaushikengineeringworks.com/what-are-the-types-of-asphalt-plants/

Index

Note: Locators in *italics* indicate figures and **bold** indicate tables.

Printed in the United States
by Baker & Taylor Publisher Services